Powering Sustainability - A Comprehensive Guide to Alternative & Renewable Energy

In a world where the demand for energy continues to rise, the quest for sustainable solutions has become paramount. "Powering Sustainability: A Comprehensive Guide to Alternative & Renewable Energy" embarks on a journey into the heart of one of humanity's most pressing challenges: the transformation of our energy landscape to ensure a greener, cleaner, and more sustainable future.

This book is a beacon for those seeking a profound understanding of alternative and renewable energy sources, their intricacies, potentials, and the pivotal role they play in addressing the environmental challenges of our time. In a rapidly evolving energy landscape, the pursuit of sustainable power sources has shifted from an option to a necessity, shaping the course of nations, industries, and individuals alike.

As the global community grapples with the consequences of fossil fuel dependency, the allure of renewable and alternative energy sources has intensified. This comprehensive guide serves as a roadmap to navigate the diverse landscape of clean energy solutions, encompassing solar, wind, hydro, geothermal, bioenergy, and more. From the fundamentals of energy conversion to cutting-edge innovations, "Powering Sustainability" unravels the science, technology, economics, and policy frameworks that underpin these transformative sources of power.

While renewable energy sources undoubtedly hold promise, the transition to a sustainable energy future is not without its challenges. This guide does not shy away from exploring the complexities of integrating renewable energy into existing infrastructures, managing intermittency, and addressing concerns related to scalability and cost-effectiveness. It seeks to empower readers with a holistic understanding, equipping them to make informed decisions and contribute to shaping a resilient and balanced energy ecosystem.

But this is not just a technical exploration. It's an ode to the power of human innovation, cooperation, and determination. Throughout these pages, you will encounter stories of communities championing solar installations, nations forging ahead with wind farms, and individuals embracing energy-efficient practices. You'll discover the pioneers who pushed the boundaries of what was once deemed impossible, catalyzing the transition to cleaner energy sources.

"Powering Sustainability" is not just a guide; it's an invitation to be part of a transformative journey. It invites policymakers, educators, industry leaders, researchers, and concerned citizens to collaborate in rewriting the narrative of energy production and consumption. It prompts us to envision a world where energy isn't a cause for environmental degradation, but a driving force for sustainability, resilience, and a more harmonious relationship with the planet.

As we embark on this enlightening expedition into the realm of alternative and renewable energy, let's grasp the profound potential that lies within our reach. With each page, we'll unveil the mechanisms, implications, and possibilities of a world powered by sustainable energy sources, igniting hope and propelling us toward a brighter, cleaner, and more prosperous future.

Introduction

- The Urgency of Sustainable Energy
- Navigating the Energy Transition
- The Role of Alternative & Renewable Energy

Chapter 1: Energy Fundamentals

- Energy Types and Conversions
- Laws of Thermodynamics in Renewable Context
- Energy Units and Measurements

Chapter 2: Solar Energy

- Photovoltaic (PV) Technology and Principles
- Solar Thermal Systems and Applications
- Solar Farms and Distributed Solar Generation

Chapter 3: Wind Energy

- Wind Turbine Technology and Aerodynamics
- Onshore and Offshore Wind Farms
- Challenges and Advances in Wind Power

Chapter 4: Hydroelectric Power

- Types of Hydropower Systems
- Run-of-River vs. Storage Hydropower
- Environmental Impacts and Sustainability

Chapter 5: Geothermal Energy

- Geothermal Heat Pumps
- Enhanced Geothermal Systems (EGS)
- Harnessing Earth's Heat for Electricity

Chapter 6: Biomass and Bioenergy

- Biomass Sources and Conversion Technologies
- Biofuels, Biogas, and Bioenergy Potential
- Environmental and Social Considerations

The Urgency of Sustainable Energy

The urgency of sustainable energy cannot be overstated. Sustainable energy refers to energy sources and practices that meet our current energy needs without compromising the ability of future generations to meet their own needs. This concept is driven by several critical factors:

1. **Climate Change**: The burning of fossil fuels like coal, oil, and natural gas for energy production is the primary driver of anthropogenic climate change. These activities release greenhouse gases, such as carbon dioxide, into the atmosphere, leading to global warming and its associated negative impacts, including rising sea levels, extreme weather events, and disruptions to ecosystems. Shifting to sustainable energy sources is essential to mitigate the worst effects of climate change.

2. **Depletion of Finite Resources**: Fossil fuels are finite resources that are being depleted at an alarming rate. As these resources become scarcer, they become more expensive to extract, leading to economic and geopolitical instability. In contrast, sustainable energy sources like solar, wind, hydroelectric, and geothermal power are renewable and virtually inexhaustible.

3. **Energy Security**: Relying heavily on fossil fuels from specific regions can lead to energy insecurity, as geopolitical tensions and supply disruptions can impact energy availability and prices. Transitioning to sustainable energy sources reduces dependence on

volatile global energy markets and enhances national and regional energy security.

4. **Air and Water Pollution**: Fossil fuel combustion releases not only greenhouse gases but also pollutants that have detrimental effects on air quality and human health. These pollutants contribute to respiratory diseases, cardiovascular issues, and other health problems. Sustainable energy technologies generate little to no air or water pollution, leading to cleaner and healthier environments.

5. **Job Creation and Economic Growth**: The transition to sustainable energy sources creates new opportunities for job growth and economic development. Renewable energy industries, such as solar and wind, have been shown to generate more jobs per unit of energy produced compared to fossil fuel industries.

6. **Technological Innovation**: Embracing sustainable energy technologies encourages innovation in various fields, including energy storage, grid management, and efficiency improvements. These innovations can lead to advancements in other industries as well.

7. **Long-Term Cost Savings**: While the initial investment in sustainable energy infrastructure may be higher, the long-term operational costs are generally lower compared to fossil fuels. Renewable energy sources have minimal fuel costs and require less maintenance, leading to cost savings over time.

8. **Global Energy Access**: Sustainable energy technologies can play a crucial role in extending access to electricity and modern energy services to underserved and remote regions, especially in developing countries. Distributed renewable energy systems can provide off-grid solutions and improve the quality of life for millions of people.

To address these urgent challenges, governments, businesses, and individuals must collaborate to accelerate the transition to sustainable energy. This involves investing in research and development, implementing supportive policies and regulations, incentivizing clean energy adoption, and raising awareness about the benefits of sustainable energy sources. The urgency of sustainable energy is not only about preserving the planet for future generations but also about creating a more stable, prosperous, and equitable world in the present.

Navigating the Energy Transition

Navigating the energy transition is a complex and multifaceted challenge that involves shifting from traditional fossil fuel-based energy systems to more sustainable and renewable alternatives. Here's a guide to help navigate this transition effectively:

1. **Set Clear Goals and Strategies**: Define clear and achievable goals for the energy transition, such as reducing greenhouse gas emissions, increasing the share of renewable energy in the energy mix, and enhancing energy efficiency. Develop comprehensive strategies that outline the steps, policies, and investments required to achieve these goals.

2. **Policy and Regulatory Frameworks**: Implement supportive policies and regulations that encourage the adoption of sustainable energy technologies. This could include incentives such as subsidies, tax breaks, feed-in tariffs, and emissions reduction targets. Clear regulatory frameworks provide stability and predictability for investors and stakeholders.

3. **Invest in Research and Development**: Allocate resources to research and develop innovative technologies that improve the efficiency, affordability, and scalability of renewable energy sources and energy storage solutions. Advances in technology will drive down costs and make renewable energy more competitive.

4. **Promote Energy Efficiency**: Energy efficiency measures can significantly reduce energy demand

and emissions. Encourage industries, businesses, and households to adopt energy-efficient practices, appliances, and building designs. Energy efficiency improvements can provide immediate benefits while transitioning to renewable energy sources.

5. **Diversify the Energy Mix**: Transitioning to a sustainable energy future involves diversifying the energy mix to include a range of renewable sources such as solar, wind, hydro, geothermal, and biomass. This reduces reliance on a single energy source and enhances energy security.

6. **Upgrade Infrastructure**: Modernize energy infrastructure, including power grids and transmission systems, to accommodate the decentralized nature of renewable energy generation. Smart grids and energy storage technologies are crucial for managing intermittent energy sources like solar and wind.

7. **Collaboration and Partnerships**: Collaboration among governments, industries, academia, and civil society is essential. Public-private partnerships can accelerate the development and deployment of sustainable energy solutions. International cooperation can facilitate knowledge sharing and technology transfer.

8. **Education and Awareness**: Raise public awareness about the benefits of sustainable energy and the importance of the energy transition. Educate consumers about energy-saving practices and the positive impact of their choices on the environment and society.

9. **Address Workforce Transition**: As the energy landscape changes, there will be shifts in employment opportunities. Invest in retraining and upskilling programs to ensure that workers in traditional energy sectors can transition to roles in the renewable energy industry.

10. **Financial Support and Investment**: Mobilize financial resources for sustainable energy projects. Encourage private sector investment by providing clear investment opportunities and a stable policy environment. Green bonds, impact investing, and venture capital can help fund innovative projects.

11. **Monitor and Evaluate Progress**: Regularly track and assess the progress of the energy transition. Use data and indicators to measure the adoption of renewable energy, emissions reductions, and economic benefits. Adjust strategies as needed based on performance.

12. **Adaptability and Flexibility**: The energy transition is a dynamic process that will require adaptation to changing technologies, market conditions, and global developments. Stay open to adjusting strategies based on new information and insights.

Successfully navigating the energy transition requires a holistic approach, combining technological innovation, policy support, financial investment, and societal engagement. The transition is an opportunity to create a cleaner, more resilient, and sustainable energy future for all.

The Role of Alternative & Renewable Energy

Alternative and renewable energy sources play a crucial role in addressing the challenges associated with traditional fossil fuel-based energy systems. These sources offer numerous benefits that contribute to sustainability, energy security, and environmental protection. Here are some key roles that alternative and renewable energy sources fulfill:

1. **Reducing Greenhouse Gas Emissions**: One of the most significant roles of renewable energy is its ability to reduce greenhouse gas emissions. Unlike fossil fuels, renewable sources like solar, wind, hydro, and geothermal power produce little to no direct emissions during operation, helping to mitigate climate change and its adverse effects.

2. **Enhancing Energy Security**: Renewable energy sources are abundant and widely distributed, reducing dependence on a few specific regions for energy supply. This enhances energy security by minimizing the geopolitical and supply risks associated with fossil fuels.

3. **Diversifying the Energy Mix**: Incorporating a diverse range of renewable energy sources diversifies the energy mix, making it more resilient to supply disruptions and price volatility. This diversity also improves grid stability and reliability.

4. **Creating Jobs and Economic Growth**: The renewable energy sector has the potential to create a

significant number of jobs across various stages, from manufacturing and installation to maintenance and operation. This contributes to local economic growth and supports sustainable development.

5. **Promoting Technological Innovation**: The pursuit of renewable energy has led to significant technological advancements in energy production, storage, and distribution. These innovations not only improve the efficiency of renewable systems but also have applications in other industries.

6. **Reducing Air and Water Pollution**: Unlike fossil fuel combustion, renewable energy sources do not release pollutants that harm air quality and human health. This improves local air quality and reduces healthcare costs associated with pollution-related illnesses.

7. **Conserving Natural Resources**: Fossil fuels are finite resources that can be depleted over time. Renewable energy sources, on the other hand, are sustainable and virtually inexhaustible, helping to preserve natural resources for future generations.

8. **Mitigating Energy Poverty**: Renewable energy technologies can be deployed in remote or underserved areas, providing access to electricity and modern energy services for communities that lack traditional power infrastructure.

9. **Empowering Individuals and Communities**: Distributed renewable energy systems, such as rooftop solar panels, enable individuals and communities to generate their own energy, reducing reliance on centralized power plants and giving them more control over their energy supply.

10. **Addressing Energy Transition Challenges**: As countries transition away from fossil fuels, renewable energy plays a pivotal role in filling the energy gap. It provides a scalable and viable alternative that can replace the energy capacity of retiring fossil fuel power

plants.

11. **Meeting International Commitments**: Many countries have committed to international agreements to reduce greenhouse gas emissions and promote sustainable development. Integrating renewable energy is a key strategy for meeting these commitments.

12. **Consumer Choice and Engagement**: Renewable energy options empower consumers to make environmentally conscious choices about their energy consumption. Programs such as net metering and feed-in tariffs incentivize individuals and businesses to adopt renewable energy systems.

13. **Long-Term Cost Savings**: While there may be upfront costs associated with installing renewable energy systems, the long-term operational costs are often lower than those of fossil fuel-based systems due to minimal fuel costs and reduced maintenance requirements.

In summary, alternative and renewable energy sources are essential components of a sustainable energy future. They offer a pathway to address climate change, improve energy security, create jobs, foster innovation, and promote overall environmental and social well-being. Transitioning to these sources is a critical step in building a resilient and prosperous global energy landscape.

Energy Types and Conversions

Energy exists in various forms, and it can be converted from one form to another through different processes. Here are some common energy types and examples of how they can be converted:

1. **Mechanical Energy**: This is the energy possessed by objects in motion or at rest. It includes both kinetic energy (energy of motion) and potential energy (energy due to position).
   - Conversion Example: When you drop a ball from a height, its potential energy is converted into kinetic energy as it falls.

2. **Heat (Thermal) Energy**: Heat energy is the energy associated with the movement of atoms and molecules within a substance. It flows from hotter to cooler regions.
   - Conversion Example: Burning fuel in an engine converts chemical energy into heat energy, which is then partially converted into mechanical energy to move a vehicle.

3. **Chemical Energy**: This energy is stored in the bonds of chemical compounds and can be released during chemical reactions.
   - Conversion Example: When you eat food, your body converts the chemical energy stored in the food into various forms of energy to perform bodily functions.

4. **Electrical Energy**: This is the energy associated with the movement of electrons through conductive

materials like wires.

- Conversion Example: Solar panels convert sunlight (radiant energy) into electrical energy through the photovoltaic effect.

5. **Radiant (Light) Energy**: Radiant energy includes visible light, radio waves, microwaves, and other forms of electromagnetic radiation.

- Conversion Example: Solar panels also convert radiant energy from the Sun into electricity through photovoltaic cells.

6. **Nuclear Energy**: This is the energy stored in the nucleus of an atom and can be released through nuclear reactions like fission or fusion.

- Conversion Example: Nuclear power plants use nuclear fission reactions to generate heat, which is then converted into electricity.

7. **Sound Energy**: Sound energy is produced by vibrations of objects and travels as waves through a medium.

- Conversion Example: Microphones convert sound energy into electrical signals, which can then be amplified and played back as sound through speakers.

8. **Kinetic Energy of Wind**: Wind energy is the kinetic energy of moving air molecules.

- Conversion Example: Wind turbines capture the kinetic energy of the wind and convert it into mechanical energy, which is further converted into electrical energy.

9. **Hydroelectric Energy**: This is the energy of moving water and is harnessed by dams and turbines.

- Conversion Example: Water flowing through a dam's turbines converts the potential energy of the water at a height into kinetic energy, which is then converted into mechanical and electrical energy.

10. **Geothermal Energy**: This is the heat energy stored within the Earth's crust.

- Conversion Example: Geothermal power plants use the heat from within the Earth to generate steam, which drives turbines to produce electricity.

These are just a few examples of the many energy types and conversions that occur in our daily lives and in various industries. Understanding these conversions is essential for designing and optimizing energy systems, improving efficiency, and making informed decisions about energy use and conservation.

Laws of Thermodynamics in Renewable Context

The laws of thermodynamics are fundamental principles that govern the behavior of energy in various systems. They are relevant in the context of renewable energy as they provide a framework for understanding energy conversion, efficiency, and limitations. Here's how the laws of thermodynamics apply to the renewable energy context:

1. **First Law of Thermodynamics (Law of Conservation of Energy)**: This law states that energy cannot be created or destroyed; it can only be transferred or converted from one form to another.

   - **Renewable Energy Context**: In the realm of renewable energy, the first law emphasizes the need to account for all forms of energy inputs and outputs in energy conversion processes. For example, in a solar panel, sunlight is converted into electricity, but the total energy remains constant, obeying the law of conservation of energy.

2. **Second Law of Thermodynamics**: The second law deals with the concept of entropy, which is a measure of the amount of energy unavailable for doing work in a system. It states that natural processes tend to increase the total entropy of a closed system over time.

   - **Renewable Energy Context**: In renewable energy systems, the second law has implications for efficiency. It sets a limit on how efficiently energy can be converted from one form to another. For instance, no energy conversion process can be 100% efficient, and some energy will always be lost as waste heat. Renewable energy technologies aim to maximize efficiency while minimizing entropy increase.

3. **Third Law of Thermodynamics**: The third law states that as the temperature of a system approaches absolute zero (0 Kelvin or -273.15°C), the entropy of the system approaches a minimum value.

   - **Renewable Energy Context**: While the third law has limited direct application to most renewable energy systems, it highlights the concept of temperature and its relationship to energy. For example, in geothermal energy, the Earth's interior heat can be harnessed for power generation based on temperature gradients.

4. **Zeroth Law of Thermodynamics**: This law defines temperature and thermal equilibrium. If two systems are separately in thermal equilibrium with a third system, they are in thermal equilibrium with each other.

   - **Renewable Energy Context**: The zeroth law is foundational for understanding heat exchange and temperature differences, which are relevant in renewable energy systems such as solar thermal collectors, geothermal systems, and heat pumps.

Understanding and applying the laws of thermodynamics is crucial when designing, analyzing, and optimizing renewable energy technologies. Engineers and researchers work within these laws to maximize energy efficiency, minimize energy losses, and develop innovative solutions for sustainable energy production and utilization.

Energy Units and Measurements

Energy is measured in various units depending on the context and the system of measurement being used. Here are some common energy units and their relationships:

1. **Joule (J)**: The joule is the standard unit of energy in the International System of Units (SI). It's defined as the amount of energy transferred when a force of one newton acts on an object to move it one meter against the force.

2. **Calorie (cal)**: The calorie is a common unit of energy used in nutrition and everyday contexts. One calorie is the amount of energy required to raise the temperature of one gram of water by one degree Celsius. The kilocalorie (kcal) is often used to describe the energy content of food.

1 kcal = 4184 J (approximately)

3. **British Thermal Unit (BTU)**: The British thermal unit is commonly used in the United States to measure energy, especially in heating and cooling systems. One BTU is the amount of energy needed to raise the temperature of one pound of water by one degree Fahrenheit.

1 BTU = 1055.06 J (approximately)

4. **Kilowatt-Hour (kWh)**: The kilowatt-hour is a unit of energy commonly used for measuring electricity consumption. It represents the amount of energy consumed when a device with a power of one kilowatt operates for one hour.

1 kWh = 3.6×10^6 J

5. **Electronvolt (eV):** The electronvolt is a unit of energy commonly used in atomic and subatomic physics. It's the energy gained or lost by an electron when it moves through an electric potential difference of one volt.

1 eV = 1.60218 × 10^-19 J

6. **Megajoule (MJ)** and Gigajoule (GJ): These are larger units of energy used to describe energy on a larger scale, such as in industrial processes or energy production.

1 MJ = 10^6 J 1 GJ = 10^9 J

When dealing with energy units, it's important to understand the relationships between them. Conversion factors allow you to convert energy from one unit to another. Additionally, power (measured in watts) is related to energy by time, as power represents the rate of energy transfer or consumption. For example, if a device has a power of 1 kW (kilowatt) and operates for 1 hour, it consumes 1 kWh (kilowatt-hour) of energy.

Having a clear understanding of energy units and measurements is essential when working with energy systems, conducting calculations, and making informed decisions about energy consumption and production.

Photovoltaic (PV) Technology and Principles

Photovoltaic (PV) technology is a method of converting sunlight directly into electricity using semiconductor materials. It's a key component of renewable energy systems and plays a crucial role in the transition to sustainable energy sources. Here are the principles and key aspects of photovoltaic technology:

1. **Photovoltaic Effect**: The photovoltaic effect is the fundamental principle behind PV technology. It's the process by which certain materials (typically semiconductors) generate an electric current when exposed to light. When photons (light particles) strike the surface of a semiconductor material, they can transfer their energy to electrons, causing them to move and create an electric current.

2. **Semiconductor Materials**: PV cells are made from semiconductor materials, usually silicon-based. Silicon has properties that make it suitable for converting light into electricity. There are two main types of silicon used in PV cells: crystalline silicon (monocrystalline and polycrystalline) and thin-film silicon.

3. **Structure of a PV Cell**: A typical PV cell consists of several layers. The most common design includes:
   - **Front Contact**: A transparent conductive layer that allows sunlight to pass through.
   - **P-N Junction**: The heart of the PV cell, where the p-type (positively charged) and n-type

(negatively charged) semiconductor layers meet. This forms a built-in electric field that helps separate charge carriers.

- **Back Contact**: Another conductive layer that collects the electrons that have been energized by sunlight.

4. **Generation of Electron-Hole Pairs**: When sunlight hits the semiconductor material, it excites electrons, creating electron-hole pairs. The electric field at the p-n junction then separates these charge carriers, pushing electrons to the n-side and holes to the p-side.

5. **Direct Current (DC) Generation**: The separated charge carriers create a voltage difference between the two sides of the p-n junction, resulting in an electric current. This current is in the form of direct current (DC).

6. **Module and Array Formation**: To generate useful amounts of electricity, individual PV cells are combined into larger units called modules or panels. Modules are then further connected in arrays to form a complete PV system. These systems can range from small rooftop installations to large solar farms.

7. **Efficiency and Performance**: PV cell efficiency refers to the ratio of usable electrical energy output to the energy of sunlight falling on the cell. Advances in PV technology focus on increasing efficiency by enhancing the cell's ability to capture and convert sunlight.

8. **Applications**: PV technology is used in a variety of applications, including residential and commercial rooftop solar panels, utility-scale solar farms, portable solar chargers, and even in space for powering satellites and spacecraft.

9. **Inverter and Grid Integration**: The electricity generated by PV panels is in the form of DC. To use it in most applications, it needs to be converted into

AC (alternating current) using an inverter. PV systems can also be integrated with the grid, allowing excess electricity to be fed back into the grid and enabling net metering.

10. **Environmental Benefits**: PV technology is a clean energy source that produces no emissions during operation. It contributes to reducing greenhouse gas emissions and reliance on fossil fuels.

Photovoltaic technology has seen significant advancements over the years, resulting in improved efficiency, reduced costs, and broader adoption. It continues to be a critical component of the global effort to transition to more sustainable and environmentally friendly energy sources.

Solar Thermal Systems and Applications

Solar thermal systems harness the heat energy from the sun to generate usable heat or electricity. Unlike photovoltaic (PV) technology, which converts sunlight directly into electricity, solar thermal systems focus on capturing and utilizing the sun's heat. Here are the principles and various applications of solar thermal systems:

1. **Principles of Solar Thermal Systems**:

Solar thermal systems operate based on the principles of collecting, concentrating, and utilizing solar heat. They typically consist of the following components:

- **Solar Collectors**: These devices capture sunlight and convert it into heat. There are different types of collectors, including flat-plate collectors and concentrating collectors like parabolic troughs and solar power towers.
- **Heat Transfer Fluid**: A fluid (usually water or a heat-transfer fluid) circulates through the collector, absorbing the heat and carrying it to the heat exchanger.
- **Heat Exchanger**: The heat exchanger transfers the collected heat from the fluid to the medium that needs to be heated, such as water for domestic use or for industrial processes.
- **Thermal Storage**: Some systems include thermal storage components, which allow captured heat to be stored for later use, such as during periods

of low sunlight.

2. **Applications of Solar Thermal Systems**:

- **Domestic Hot Water**: Solar water heating systems use solar collectors to heat water for household use. These systems can be installed on rooftops and significantly reduce the energy required for water heating.
- **Space Heating**: Solar thermal systems can provide heat for space heating in buildings. This can be achieved through radiant heating systems that distribute heated fluid through floors or walls.
- **Solar Cooking**: Solar cookers and solar ovens use solar thermal energy to cook food. They are particularly useful in areas with limited access to traditional cooking fuels.
- **Industrial Processes**: Solar thermal systems are used in various industrial processes that require heat, such as drying, sterilization, and certain manufacturing processes.
- **Desalination**: Solar thermal desalination systems use solar heat to evaporate and condense water, producing fresh water from saltwater sources.
- **Power Generation**: Concentrated solar power (CSP) plants use mirrors or lenses to concentrate sunlight onto a receiver, producing high-temperature heat that is then used to generate electricity through conventional steam turbines.
- **Solar Cooling**: Solar thermal systems can drive absorption chillers or desiccant systems, providing cooling for buildings using solar heat.
- **Agricultural and Aquaculture Applications**: Solar thermal systems can be used

for greenhouse heating, drying crops, and maintaining optimal temperatures in aquaculture systems.

3. **Advantages of Solar Thermal Systems**:
   - Solar thermal systems can achieve higher temperatures than solar photovoltaic systems, making them suitable for applications that require heat at elevated temperatures.
   - They can provide reliable and consistent heat output, especially when equipped with thermal storage.
   - Solar thermal technology has been in use for decades and is well-established in various applications.

4. **Challenges and Considerations**:
   - Solar thermal systems are generally more complex to design and install compared to photovoltaic systems.
   - The efficiency of solar thermal systems can be influenced by factors such as weather conditions and system design.
   - Some applications may require backup systems for periods of low sunlight.

Overall, solar thermal systems offer a versatile and sustainable solution for harnessing solar energy to meet various heating and energy needs across residential, commercial, industrial, and agricultural sectors.

Solar Farms and Distributed Solar Generation

Solar energy is harnessed through two primary approaches: solar farms (also known as utility-scale solar) and distributed solar generation. Both play a crucial role in meeting energy needs and transitioning to more sustainable energy sources. Here's an overview of each approach:

1. **Solar Farms (Utility-Scale Solar):**

Solar farms are large installations that generate electricity from sunlight on a massive scale. They are often located in areas with ample sunlight and available land. Here are the key aspects of solar farms:

- **Size and Capacity**: Solar farms can cover several acres or even hundreds of acres. They consist of multiple rows of solar panels that can collectively generate large amounts of electricity.
- **Efficiency**: Solar farms use economies of scale to achieve higher efficiency in energy generation. The larger surface area of panels allows for more sunlight to be captured and converted into electricity.
- **Grid Integration**: Solar farms are typically connected to the electricity grid. Excess electricity generated during sunny periods can be fed back into the grid, while power is drawn from the grid when sunlight is insufficient.
- **Applications**: Solar farms contribute to the overall energy supply of a region or country. They

play a significant role in reducing greenhouse gas emissions and diversifying the energy mix.

- **Location**: Solar farms are often established in areas with high solar irradiance, such as deserts or open rural landscapes. The choice of location takes into account factors like land availability, solar resource, and proximity to transmission infrastructure.

2. **Distributed Solar Generation**:

Distributed solar generation refers to smaller-scale solar installations located closer to the point of energy consumption. It encompasses residential, commercial, and industrial solar systems. Here's what you need to know about distributed solar:

- **Size and Capacity**: Distributed solar systems vary in size, from rooftop installations on individual homes to larger arrays on commercial buildings or industrial facilities.
- **On-Site Consumption**: Distributed solar systems generate electricity that can be used on-site, reducing the need to draw energy from the grid. Excess energy can be fed back into the grid through net metering programs.
- **Energy Independence**: Distributed solar generation empowers individuals, businesses, and communities to generate their own electricity and reduce reliance on centralized power sources.
- **Location**: Distributed solar can be integrated into existing infrastructure, such as rooftops and parking structures, maximizing space utilization and minimizing land use.
- **Grid Support**: Distributed solar systems can enhance grid stability by reducing peak demand and providing localized energy generation.
- **Resilience**: Distributed solar can provide backup

power during grid outages when coupled with energy storage systems, increasing resilience against power disruptions.

Both solar farms and distributed solar generation contribute to the growth of renewable energy capacity. They address different scales of energy demand and offer solutions for transitioning to cleaner energy sources. Collectively, they help reduce greenhouse gas emissions, improve energy security, and create a more sustainable energy landscape.

Wind Turbine Technology and Aerodynamics

Wind turbine technology harnesses the kinetic energy of wind to generate electricity. These devices have become a key component of renewable energy systems and play a significant role in transitioning to cleaner energy sources. Let's delve into the principles of wind turbine technology and the aerodynamics that drive their operation:

1. **Principles of Wind Turbine Operation**:
Wind turbines work based on the principles of aerodynamics and energy conversion. Here's an overview of how they operate:
   - **Blade Rotation**: Wind energy causes the turbine's blades to rotate. The blades are designed to capture as much wind energy as possible.
   - **Mechanical Energy**: The rotational motion of the blades is converted into mechanical energy, which is transmitted to a generator through a shaft.
   - **Generator Conversion**: The generator converts the mechanical energy into electrical energy by using the motion to induce a magnetic field and generate an electric current.

2. **Aerodynamics of Wind Turbines**:
The aerodynamics of wind turbines are crucial to their efficiency and performance. Understanding how air flows over and around the blades helps engineers design more effective turbines. Some key aerodynamic concepts

include:

- **Lift and Drag**: Wind turbine blades are designed similarly to aircraft wings. They generate lift and drag forces as air flows over them. Lift is the upward force that provides the blade with its rotational motion, while drag is the resistance force that opposes the motion.
- **Angle of Attack**: The angle at which the blade meets the oncoming wind is called the angle of attack. Adjusting this angle allows engineers to optimize lift and minimize drag.
- **Tip Speed Ratio**: The tip speed ratio is the ratio of the speed of the blade tips to the wind speed. It helps determine the most efficient rotational speed of the blades for a given wind speed.
- **Turbulence and Stall**: Turbulent air and high wind speeds can cause the blades to stall, reducing their efficiency. Turbine designs often incorporate features to minimize stall and handle turbulent conditions.
- **Yaw Control**: Yaw control involves adjusting the orientation of the turbine to face the wind. This ensures that the blades capture the maximum energy from the wind direction.

3. **Types of Wind Turbines**:

There are two main types of wind turbines: horizontal-axis wind turbines (HAWTs) and vertical-axis wind turbines (VAWTs).

- **HAWTs**: These are the most common type and have a horizontal rotor shaft, with the blades rotating parallel to the ground. They are typically more efficient at higher wind speeds.
- **VAWTs**: These have a vertical rotor shaft and blades that rotate around it. VAWTs are often used in urban environments and areas with variable wind directions.

4. **Advancements in Wind Turbine Technology**:
Wind turbine technology has evolved significantly, leading to larger, more efficient, and more reliable turbines. Some advancements include:

- **Increased Blade Length**: Longer blades capture more wind energy and improve efficiency.
- **Advanced Materials**: Lightweight and durable materials enhance blade design and overall turbine performance.
- **Control Systems**: Advanced control systems optimize blade pitch, yaw, and turbine speed for maximum energy capture and stability.
- **Offshore Wind Turbines**: Installing wind turbines offshore captures stronger and more consistent wind, making them increasingly popular for large-scale energy production.

Wind turbine technology continues to advance, contributing to the growth of renewable energy capacity and reducing reliance on fossil fuels.

Onshore and Offshore Wind Farms

Onshore and offshore wind farms are installations that harness wind energy to generate electricity. They are both important components of the renewable energy landscape and play a crucial role in reducing greenhouse gas emissions and transitioning to cleaner energy sources. Here's a comparison of onshore and offshore wind farms:

Onshore Wind Farms:

1. **Location**: Onshore wind farms are located on land. They are typically situated in areas with favorable wind conditions, such as open plains, ridges, or coastal areas.

2. **Advantages**:
   - Onshore wind farms are generally easier and less costly to install and maintain compared to offshore wind farms.
   - They have shorter transmission distances, which can reduce energy transmission losses.
   - Onshore wind farms can often benefit from existing infrastructure, such as roads and power lines.

3. **Challenges**:
   - They may face land-use conflicts, particularly in areas with competing land uses or concerns about visual impact.
   - Wind speeds can be lower and more variable compared to offshore locations, affecting energy production.

4. **Environmental Considerations**:

- Onshore wind farms can have some impact on local ecosystems, bird populations, and land use.
- Careful site selection and environmental impact assessments are important to minimize negative effects.

Offshore Wind Farms:

1. **Location**: Offshore wind farms are located in bodies of water, typically in coastal areas or out at sea. They take advantage of stronger and more consistent wind speeds offshore.
2. **Advantages**:
    - Offshore wind farms benefit from higher and more consistent wind speeds, leading to potentially higher energy production.
    - They have less visual impact and fewer land-use conflicts compared to onshore installations.
    - Offshore wind farms can be larger in scale and contribute significantly to national energy capacity.
3. **Challenges**:
    - Offshore wind farms require more complex engineering and installation due to the harsh marine environment and logistical challenges.
    - They have higher construction and maintenance costs compared to onshore installations.
    - Transmission infrastructure to connect offshore wind farms to the grid can be more expensive.
4. **Environmental Considerations**:
    - Offshore wind farms can have less impact on

terrestrial ecosystems, but they can still affect marine habitats, fish populations, and bird migration routes.
- Proper environmental impact assessments and monitoring are crucial to minimize negative effects.

5. **Advancements**:
- Advancements in offshore wind technology have led to the development of larger turbines and floating platforms, enabling the deployment of wind farms in deeper waters.

Both onshore and offshore wind farms contribute to the expansion of renewable energy capacity and reducing reliance on fossil fuels. The choice between onshore and offshore installations depends on factors such as wind resource, land availability, environmental considerations, and economic feasibility. Many regions are embracing both types of wind farms to create a diverse and resilient renewable energy portfolio.

Challenges and Advances in Wind Power

Wind power has made significant strides as a clean and renewable energy source, but it still faces various challenges that researchers, engineers, and policymakers are actively working to address. At the same time, there have been noteworthy advances in wind power technology that are helping to overcome these challenges. Here's an overview of some key challenges and advances in wind power:

Challenges:

1. **Intermittency and Variability**: Wind power generation is subject to the variability of wind speeds, which can lead to fluctuations in electricity output. This intermittency poses challenges for grid integration and stability.
2. **Grid Integration**: Integrating large amounts of wind power into the grid requires advanced grid management systems to balance supply and demand and ensure stability.
3. **Land Use and Siting**: Onshore wind farms require substantial land area, which can lead to conflicts with other land uses and concerns about visual impacts. Proper siting and land use planning are essential.
4. **Environmental Impact**: Wind farms can have environmental impacts on local ecosystems, bird populations, and wildlife habitats. Balancing renewable energy goals with environmental protection is important.
5. **Resource Limitations**: Wind power availability depends on suitable wind resources, which may not

be equally distributed across all regions. This can limit the feasibility of wind power in certain areas.

Advances:

1. **Increased Efficiency and Capacity**: Advances in turbine design, materials, and engineering have led to larger and more efficient wind turbines with higher energy-capturing capabilities.
2. **Floating Offshore Wind Farms**: Floating platforms enable the deployment of offshore wind farms in deeper waters, expanding the potential for offshore wind power generation.
3. **Advanced Control Systems**: Smart control systems optimize wind turbine operation by adjusting blade pitch, yaw, and speed based on real-time wind conditions, improving efficiency and stability.
4. **Energy Storage Integration**: Pairing wind power with energy storage technologies, such as batteries, helps mitigate intermittency issues by storing excess energy and releasing it when needed.
5. **Hybrid Systems**: Combining wind power with other renewable sources like solar or hydropower in hybrid systems provides more consistent energy generation and grid stability.
6. **Machine Learning and Data Analytics**: Advanced data analytics and machine learning techniques are being used to predict wind patterns, optimize turbine performance, and improve maintenance strategies.
7. **Decentralized and Community-Based Projects**: Community wind projects allow local communities to participate in and benefit from wind power generation, promoting local ownership and support.
8. **Policy Support and Market Incentives**: Favorable policies, subsidies, and market incentives in many regions have accelerated the growth of wind power

installations and technology development.

9. **Offshore Wind Growth**: Offshore wind technology has seen significant growth, with larger turbines and improved installation techniques, leading to more efficient and cost-effective projects.

Addressing the challenges and leveraging these advances is essential for the continued growth of wind power as a reliable and sustainable energy source. As technology continues to evolve, wind power is likely to play an increasingly prominent role in global efforts to mitigate climate change and transition to cleaner energy systems.

Types of Hydropower Systems

Hydropower is a renewable energy technology that harnesses the energy of flowing water to generate electricity. There are several types of hydropower systems, each with its own characteristics and applications. Here are some common types:

1. **Conventional Hydropower**:

Conventional hydropower systems typically involve the construction of dams and reservoirs to store water. When electricity is needed, water is released from the reservoir, flows through turbines, and generates electricity. There are two main types of conventional hydropower:

- **Run-of-River Hydropower**: In this system, the natural flow of a river is used to generate electricity without significantly altering the river's flow or diverting large amounts of water. Run-of-river systems usually have smaller environmental impacts compared to reservoir-based systems.

- **Reservoir Hydropower**: This type involves the construction of a dam and reservoir to store water. Water is released from the reservoir to generate electricity during periods of high demand. Reservoir systems can provide a more consistent energy output and offer opportunities for energy storage.

2. **Pumped Storage Hydropower**:

Pumped storage is a type of hydropower that acts as a form of energy storage. It involves two reservoirs located at different elevations. During periods of excess electricity supply (low demand), surplus energy is used to pump water from the lower reservoir to the upper reservoir. When electricity demand is high, water is released from the upper reservoir to the lower reservoir through turbines, generating electricity. Pumped storage facilities help stabilize the grid by providing a means of quickly responding to changes in demand.

3. **Small Hydropower**:

Small hydropower systems, also known as micro-hydropower or mini-hydropower, are designed for lower power outputs and are often used in rural or remote areas. They can utilize the natural flow of streams and rivers or divert water through small turbines to generate electricity for local communities or specific applications.

4. **Tidal Hydropower**:

Tidal hydropower captures energy from the rise and fall of ocean tides. It involves building structures, such as tidal barrages or tidal stream turbines, in coastal areas where tidal currents are strong. These structures harness the kinetic energy of tidal flows to generate electricity.

5. **Run-of-River Diversion**:

This type of system diverts a portion of a river's flow through a canal or pipeline, passing it through turbines to generate electricity before returning the water to the river downstream. It aims to minimize the environmental impact compared to large reservoir-based systems.

6. **In-Stream Hydropower**:

In-stream or free-flow hydropower systems involve placing turbines directly in flowing rivers or streams without the need for dams or reservoirs. These systems generate electricity from the kinetic energy of the flowing water.

Each type of hydropower system has its advantages and challenges, and the choice of system depends on factors such as water availability, environmental considerations, energy demand, and economic feasibility. Hydropower continues to be a significant source of renewable energy, contributing to global efforts to reduce greenhouse gas emissions and transition to cleaner energy sources.

Run-of-River vs. Storage Hydropower

"Run-of-river" and "storage" are two different approaches to hydropower generation, each with its own characteristics, advantages, and challenges. Here's a comparison of run-of-river hydropower and storage hydropower:

Run-of-River Hydropower:

1. **Operation**: Run-of-river hydropower systems utilize the natural flow of a river or stream to generate electricity without significantly altering the watercourse. Water is diverted from the river through a canal or penstock and directed to turbines. After passing through the turbines, the water is returned to the river downstream.

2. **Advantages**:
   - Minimal Environmental Impact: Run-of-river systems usually have less environmental impact compared to storage systems because they don't require large reservoirs or significant changes to river flows.
   - Faster Deployment: These systems can be quicker to deploy since they do not involve the construction of large dams and reservoirs.
   - Smaller Land Footprint: Run-of-river projects typically require less land compared to storage hydropower projects.

3. **Challenges**:

- Seasonal Variability: The energy output of run-of-river systems can be influenced by seasonal variations in river flow, affecting their reliability.
- Limited Energy Storage: Run-of-river systems have limited capacity to store excess energy for times when demand is higher than river flow.

Storage Hydropower:

1. **Operation**: Storage hydropower, also known as reservoir hydropower, involves building dams and reservoirs to store water. Water is released from the reservoir to flow through turbines and generate electricity when demand is high.
2. **Advantages**:
   - Energy Storage: Storage hydropower offers the ability to store water in reservoirs, allowing for electricity generation on-demand, which helps stabilize the grid and respond to changing energy demand.
   - Consistent Energy Output: These systems can provide a more consistent energy output compared to run-of-river systems since they are not solely dependent on river flow.
3. **Challenges**:
   - Environmental Impact: The construction of dams and reservoirs can have significant environmental impacts, including habitat disruption and altered river ecosystems.
   - Land and Space Requirements: Storage hydropower projects often require more land for reservoirs and can have larger land footprints.
   - Longer Deployment Time: The construction

of dams and reservoirs can extend the project timeline compared to run-of-river systems.

Considerations:

The choice between run-of-river and storage hydropower depends on factors such as the available water resources, environmental considerations, energy demand patterns, and project feasibility. Some regions may opt for run-of-river systems to minimize environmental impact, while others may choose storage systems for their energy storage capabilities. Both approaches contribute to clean energy generation and can play a role in achieving renewable energy goals.

Environmental Impacts and Sustainability

Hydropower, while a renewable energy source, can have both positive and negative environmental impacts. Understanding these impacts is crucial for making informed decisions about the sustainability of hydropower projects. Here's an overview of the environmental impacts and considerations associated with hydropower:

Positive Environmental Impacts:

1. **Greenhouse Gas Emissions Reduction**: Hydropower generates electricity without direct greenhouse gas emissions, contributing to reduced reliance on fossil fuels and mitigating climate change.
2. **Air and Water Quality Improvement**: Hydropower produces minimal air pollutants and does not emit pollutants that contribute to air quality degradation. It also has a smaller water footprint compared to fossil fuel-based power generation.
3. **Water Management**: Hydropower facilities can help regulate water flow and manage flooding, particularly during periods of heavy rainfall or snowmelt.
4. **Enhanced Ecosystems**: In some cases, reservoirs created by hydropower projects can provide new habitats for fish and other aquatic species, enhancing local biodiversity.

Negative Environmental Impacts:

1. **Habitat Disruption**: The construction of dams and reservoirs can disrupt natural river ecosystems, alter fish migration routes, and impact aquatic habitats.
2. **Fish Population Decline**: Dams can prevent fish from migrating upstream to spawn, leading to declines in fish populations. Fish passage systems are often required to mitigate this impact.
3. **Altered River Flow**: The alteration of river flows can affect downstream ecosystems, sediment transport, and nutrient cycling, leading to changes in river dynamics.
4. **Methane Emissions**: In reservoirs, organic matter can decompose and release methane, a potent greenhouse gas. Methane emissions from reservoirs can contribute to climate change.
5. **Sedimentation**: Reservoirs can trap sediment that would naturally flow downstream. This can lead to downstream erosion, loss of habitats, and changes in water quality.
6. **Social and Cultural Impacts**: Large-scale hydropower projects can displace communities and impact local cultures and livelihoods. Addressing these social considerations is important for project sustainability.

Sustainable Hydropower Practices:

1. **Site Selection**: Careful site selection is critical to minimize environmental impacts. Evaluating potential impacts on ecosystems, fisheries, and communities is essential.
2. **Fish Passage and Migration**: Implementing fish-friendly technologies, such as fish ladders and fish lifts, can help maintain fish populations and support river ecosystems.
3. **Environmental Mitigation and Restoration**: Integrating environmental mitigation measures and

restoration efforts into project design can help offset negative impacts.

4. **Adaptive Management**: Monitoring and adjusting hydropower operations based on environmental data and feedback can minimize negative effects.

5. **Small-Scale and Run-of-River Projects**: Smaller, run-of-river projects often have fewer environmental impacts compared to large storage-based projects.

6. **Public Engagement**: Engaging with local communities, stakeholders, and indigenous groups throughout the project lifecycle can lead to more sustainable outcomes.

Sustainable hydropower development involves careful planning, collaboration, and balancing energy generation with environmental and social considerations. As with any energy source, the goal is to minimize negative impacts while maximizing the benefits of clean and reliable electricity generation.

Geothermal Heat Pumps

Geothermal heat pumps, also known as ground-source heat pumps (GSHPs), are highly efficient heating and cooling systems that use the natural thermal energy stored in the earth to regulate indoor temperatures. These systems take advantage of the relatively constant temperature of the ground just a few feet below the Earth's surface. Here's how geothermal heat pumps work and their key features:

How Geothermal Heat Pumps Work:

1. **Heat Exchange**: Geothermal heat pumps use a closed-loop system of pipes buried in the ground, either horizontally or vertically. These pipes are filled with a heat transfer fluid, usually water mixed with antifreeze.
2. **Heat Absorption**: In heating mode, the fluid absorbs heat from the ground through the pipes. The ground's temperature is relatively stable throughout the year, typically warmer than the outdoor air in winter.
3. **Heat Pump**: The heat transfer fluid circulates through the pipes and enters the heat pump unit located inside the building. The heat pump contains a compressor, a heat exchanger, and a refrigerant.
4. **Heat Extraction**: The heat pump's compressor pressurizes the refrigerant, causing it to evaporate and absorb the heat from the heat transfer fluid. The heated refrigerant vapor is then passed through a heat exchanger, transferring its heat to the indoor air.
5. **Distribution**: The heated air is distributed throughout the building using a fan or a forced-air system. In

cooling mode, the process is reversed, and the heat pump transfers excess heat from the indoor air to the heat transfer fluid, which is then expelled into the cooler ground.

Key Features and Benefits:

1. **Energy Efficiency**: Geothermal heat pumps are highly efficient, with a high coefficient of performance (COP). They use much less energy than traditional heating and cooling systems.
2. **Constant Temperatures**: The ground temperature is relatively stable throughout the year, providing a consistent heat source or heat sink for the heat pump.
3. **Low Operating Costs**: Due to their efficiency, geothermal heat pumps can result in lower energy bills and long-term savings despite higher upfront installation costs.
4. **Environmental Impact**: Geothermal systems use renewable energy and produce fewer greenhouse gas emissions compared to conventional heating and cooling systems.
5. **Space Savings**: Geothermal systems do not require outdoor units like air-source heat pumps or air conditioners, saving outdoor space.
6. **Quiet Operation**: Geothermal systems are typically quieter than traditional HVAC systems since they lack noisy outdoor units.
7. **Durability**: The components of geothermal systems are underground or indoors, increasing their lifespan and reducing exposure to harsh weather conditions.
8. **Flexible Installation**: Geothermal systems can be installed horizontally or vertically, making them suitable for various property sizes and configurations.
9. **Incentives**: Many regions offer incentives and tax credits for installing geothermal heat pump systems,

which can offset the initial installation costs.

While geothermal heat pumps offer numerous benefits, their successful implementation requires proper system design, installation, and maintenance. Factors such as site conditions, geological considerations, and the specific heating and cooling needs of the building must be carefully assessed to ensure optimal performance.

Enhanced Geothermal Systems (EGS)

Enhanced Geothermal Systems (EGS) are a type of geothermal energy technology designed to extract heat from the Earth's subsurface even in areas where traditional geothermal resources are not naturally abundant. EGS is a groundbreaking approach that aims to create artificial reservoirs by stimulating subsurface rock formations, allowing for the production of geothermal energy on a much larger scale. Here's how Enhanced Geothermal Systems work and their significance:

How Enhanced Geothermal Systems Work:

1. **Reservoir Creation**: EGS involves creating an underground reservoir where heat can be extracted. This is done by drilling deep into hot rock formations, typically at depths of several kilometers.
2. **Stimulation**: Once the well is drilled, water is injected into the hot rock at high pressure to create fractures or pathways for heat transfer. This process is known as hydraulic fracturing or "fracking."
3. **Heat Extraction**: The injected water heats up as it comes into contact with the hot rock. The heated water is then pumped back to the surface through a separate production well.
4. **Electricity Generation**: At the surface, the heated water is used to generate steam that drives a turbine connected to a generator, producing electricity.
5. **Reinjection**: After heat extraction, the cooled water can be reinjected into the reservoir to maintain reservoir pressure and stimulate further heat exchange.

Significance and Advantages:

1. **Ubiquitous Resource**: EGS expands the potential for geothermal energy production to regions where natural geothermal resources are limited. This increases the availability of renewable energy sources globally.
2. **Baseload Power**: EGS systems can provide baseload power, which means they can operate consistently, day and night, providing a stable and reliable energy source.
3. **Low Environmental Impact**: EGS produces minimal greenhouse gas emissions compared to fossil fuels. Additionally, it doesn't rely on external fuel supply, reducing transportation and associated emissions.
4. **Energy Security**: EGS contributes to energy diversification and security by providing a local and sustainable energy source.
5. **Scalability**: EGS can be deployed at various scales, from small power plants to larger installations, allowing for flexibility in meeting energy demand.

Challenges and Considerations:

1. **Technical Challenges**: Creating and maintaining fractures in deep rock formations can be technically challenging and costly. Ensuring efficient and reliable heat transfer from the rock to the fluid is critical.
2. **Seismicity**: Injecting fluids into the Earth's crust can induce seismic activity, which needs careful monitoring and management.
3. **Water Usage**: EGS requires substantial amounts of water for injection and cooling. Managing water resources is important for sustainable operations.
4. **Economic Viability**: The initial investment in drilling and reservoir stimulation can be high. EGS projects

require careful economic analysis to determine their viability.

5. **Regulatory and Public Acceptance**: EGS operations involve hydraulic fracturing, which can raise regulatory and public acceptance concerns related to environmental and seismic impacts.

EGS technology holds the potential to significantly expand the use of geothermal energy for electricity generation and heating. As research and development continue, EGS could become an important contributor to global efforts to reduce carbon emissions and transition to more sustainable energy sources.

Harnessing Earth's Heat for Electricity

Harnessing the Earth's heat for electricity involves tapping into geothermal energy sources, which naturally exist within the Earth's subsurface. Geothermal energy is a reliable and renewable resource that can be used for both direct heating applications and electricity generation. Here's an overview of how geothermal energy is harnessed for electricity:

1. **Natural Geothermal Heat**: The Earth's interior contains a significant amount of heat, originating from the planet's formation and radioactive decay processes. This heat gradually moves towards the surface, creating temperature variations in the subsurface.

2. **Geothermal Reservoirs**: Certain areas of the Earth's crust contain higher concentrations of heat due to the proximity of molten rock (magma) or hot rock formations. These regions are known as geothermal reservoirs.

3. **Direct Use and Heat Pump Systems**: In some cases, geothermal heat can be used directly for space heating, greenhouse cultivation, industrial processes, and even spa bathing. Heat pump systems are also used to extract heat from shallow ground or groundwater for space heating and cooling.

4. **Electricity Generation**: For electricity generation, the focus is on high-temperature geothermal resources. These are typically found in regions with active

tectonic plate boundaries, such as geothermal hotspots, volcanic areas, and geysers. The process involves several steps:

- **Exploration and Drilling**: Exploration identifies potential geothermal reservoirs. Deep wells are drilled into these reservoirs to reach the hot rock formations.
- **Fluid Circulation**: Water or other fluids are injected into the wells, circulating through the hot rock, and absorbing heat from the subsurface.
- **Steam Production**: The heated fluid returns to the surface as steam or hot water due to the high temperatures in the reservoir. This steam can be used directly for electricity generation.
- **Power Generation**: The steam is used to drive a turbine connected to a generator. The turbine's mechanical energy is converted into electricity.
- **Reinjection**: After steam is used, the cooled fluid is reinjected into the reservoir to maintain pressure and continue the heat exchange process.

5. **Binary Cycle Power Plants**: In areas with lower temperature geothermal resources, binary cycle power plants are used. These plants transfer heat from the geothermal fluid to another fluid with a lower boiling point. The second fluid vaporizes and drives a turbine connected to a generator.

6. **Environmental Benefits**: Geothermal power generation produces minimal greenhouse gas emissions compared to fossil fuels. It offers a stable and reliable source of electricity, contributing to energy security and reducing dependence on fossil fuels.

7. **Challenges and Considerations**: Geothermal power generation is limited to areas with accessible geothermal reservoirs. Proper reservoir management is crucial to maintain long-term sustainability. Additionally, drilling and infrastructure costs can be high.

Harnessing the Earth's heat for electricity offers a valuable way to diversify the energy mix, reduce carbon emissions, and provide consistent power generation. It plays a role in meeting global energy demands while contributing to a more sustainable future.

Biomass Sources and Conversion Technologies

Biomass is organic material derived from plants, animals, and microorganisms. It can be used as a renewable energy source through various conversion technologies that extract energy from the chemical bonds within biomass. Here's an overview of biomass sources and the technologies used for biomass conversion:

Biomass Sources:

1. **Woody Biomass**: This includes trees, branches, wood chips, and wood pellets. It's commonly used for heating and electricity generation.
2. **Agricultural Residues**: Crop residues such as corn stalks, wheat straw, and rice husks are used as biomass feedstock.
3. **Energy Crops**: Certain crops, such as switchgrass and miscanthus, are grown specifically for energy production due to their high biomass yield.
4. **Animal Waste**: Manure from livestock and poultry can be used for biogas production.
5. **Algae**: Algae can be cultivated to produce biomass for biofuels and other products.
6. **Municipal Solid Waste**: Organic waste from households and industries can be used as biomass feedstock.

Biomass Conversion Technologies:

1. **Combustion**: Biomass can be burned directly to produce heat or electricity. It's commonly used in residential heating, industrial boilers, and power plants.
2. **Gasification**: In gasification, biomass is heated in a low-oxygen environment to produce syngas (a mixture of carbon monoxide, hydrogen, and other gases). Syngas can be used for electricity generation, heating, or converted into biofuels.
3. **Pyrolysis**: Pyrolysis involves heating biomass in the absence of oxygen to produce bio-oil, char, and gases. Bio-oil can be further refined into biofuels.
4. **Anaerobic Digestion**: This process involves breaking down organic materials in the absence of oxygen to produce biogas, primarily composed of methane and carbon dioxide. Biogas can be used for heat, electricity, or as a vehicle fuel.
5. **Fermentation**: Biomass can be converted into biofuels through fermentation. For example, sugars from crops can be fermented to produce bioethanol, commonly used as a gasoline additive.
6. **Biochemical Conversion**: Enzymes or microorganisms are used to break down biomass into sugars, which can then be fermented into biofuels or other products.
7. **Cofiring**: Biomass can be mixed with coal and burned in existing coal-fired power plants, reducing the carbon emissions from these plants.
8. **Combined Heat and Power (CHP)**: Also known as cogeneration, CHP systems use biomass to generate both heat and electricity, improving overall energy efficiency.

Benefits and Considerations:

Benefits:

- Biomass is a renewable energy source that can help

reduce reliance on fossil fuels.

- It can provide a reliable and dispatchable source of energy.
- Utilizing biomass waste reduces landfill waste and associated emissions.
- Biomass conversion can create jobs in agriculture, forestry, and related industries.

Considerations:

- Sustainable sourcing is important to prevent deforestation and habitat destruction.
- Biomass feedstock availability can be influenced by seasonal variations and crop cycles.
- Efficient conversion requires proper technologies and management to minimize emissions.
- Some conversion methods can require significant investment and operational costs.

Biomass offers a versatile and flexible energy source that can play a role in transitioning to a more sustainable energy mix. Careful consideration of feedstock selection, conversion technology, and environmental impacts is essential for maximizing its benefits.

Biofuels, Biogas, and Bioenergy Potential

Biofuels, biogas, and bioenergy are important components of the renewable energy landscape, derived from organic materials and waste. They offer alternatives to fossil fuels and contribute to reducing greenhouse gas emissions. Here's an overview of biofuels, biogas, and their potential in the context of bioenergy:

Biofuels:

Biofuels are liquid or gaseous fuels derived from biomass sources, such as crops, agricultural residues, and waste materials. They can be used as substitutes for gasoline, diesel, and aviation fuels. There are two main types of biofuels:

1. **Bioethanol**: Bioethanol is an alcohol-based biofuel primarily produced through the fermentation of sugars from crops like corn, sugarcane, and cellulosic biomass. It is commonly used as a gasoline additive to reduce emissions.
2. **Biodiesel**: Biodiesel is produced from vegetable oils, animal fats, or recycled cooking oil through a chemical process called transesterification. Biodiesel can be blended with conventional diesel and used in diesel engines.

Biogas:

Biogas is a mixture of methane and carbon dioxide produced by the anaerobic digestion of organic materials such as agricultural waste, animal manure, and sewage. It can be used for electricity generation, heating, and even as a vehicle fuel. Biogas is a versatile energy source that also helps in waste management by utilizing organic waste.

Bioenergy Potential:

1. **Reducing Emissions**: Biofuels and biogas have the potential to significantly reduce greenhouse gas emissions compared to fossil fuels, contributing to climate change mitigation.
2. **Renewable Energy**: Bioenergy sources are renewable and can be continuously produced from organic materials, as long as sustainable practices are followed.
3. **Waste Utilization**: Bioenergy can help convert organic waste, agricultural residues, and other biomass materials into useful energy, reducing the environmental impact of waste disposal.
4. **Energy Security**: Bioenergy can enhance energy security by providing locally sourced and decentralized energy production, reducing reliance on imported fossil fuels.
5. **Economic Opportunities**: The bioenergy sector can create jobs in agriculture, forestry, and renewable energy industries, supporting rural economies.

Challenges and Considerations:

1. **Land Use Competition**: There's a concern that biofuel production might compete with food production, leading to increased food prices and deforestation. Sustainable land management and careful crop selection are necessary.
2. **Resource Availability**: The availability of biomass resources varies geographically and seasonally, which can impact the reliability of bioenergy production.
3. **Technology Development**: Advancements in

conversion technologies, such as cellulosic biofuel production and efficient biogas digestion systems, are needed to improve overall efficiency.

4. **Environmental Impact**: Unsustainable bioenergy practices can lead to habitat destruction, soil degradation, and water resource depletion. Responsible sourcing and land management are crucial.

5. **Policy Support**: Government policies and incentives play a vital role in promoting the growth of the bioenergy sector and ensuring its sustainability.

Biofuels, biogas, and bioenergy have the potential to play a significant role in transitioning to a more sustainable energy system. However, achieving their potential while addressing environmental and social challenges requires careful planning, technological innovation, and sustainable practices.

Environmental and Social Considerations

Environmental and social considerations are paramount when developing and deploying renewable energy technologies. While these technologies offer the potential to reduce greenhouse gas emissions and promote sustainability, their implementation can also have various impacts on ecosystems, communities, and societies. Here's a closer look at the environmental and social considerations associated with renewable energy:

Environmental Considerations:

1. **Habitat Disruption**: The construction of renewable energy infrastructure, such as wind turbines, solar panels, and hydropower dams, can disrupt natural habitats, affecting wildlife and ecosystems.
2. **Land Use**: The allocation of land for renewable energy projects might compete with other land uses, such as agriculture, forestry, and conservation.
3. **Water Usage**: Certain renewable energy technologies, like hydropower and bioenergy, can require significant water resources for operation and cooling.
4. **Visual and Aesthetic Impact**: Large-scale renewable energy installations can alter landscapes and potentially impact the aesthetic value of natural areas.
5. **Resource Extraction**: The production of materials needed for renewable technologies, such as rare earth metals for solar panels and wind turbines, can have environmental impacts during extraction and processing.
6. **End-of-Life Management**: Proper disposal and

recycling of renewable energy equipment, such as solar panels and batteries, are important to minimize waste and pollution.

Social Considerations:

1. **Community Engagement**: Early and meaningful engagement with local communities is crucial to address concerns, provide information, and involve stakeholders in decision-making processes.
2. **Cultural Heritage**: Renewable energy projects should consider the cultural and historical significance of the land to indigenous and local communities.
3. **Health and Safety**: Potential health and safety impacts of renewable energy projects on nearby communities, such as noise pollution or electromagnetic fields, should be assessed.
4. **Job Creation and Local Economy**: Renewable energy projects can create job opportunities and stimulate local economies. Ensuring that local communities benefit from these opportunities is important.
5. **Land Rights and Displacement**: The development of renewable energy projects can lead to land acquisition and displacement of communities. Ensuring fair compensation and safeguarding land rights is essential.
6. **Energy Access**: Renewable energy projects can contribute to expanding access to electricity in underserved regions, improving living conditions and opportunities.
7. **Equity and Accessibility**: Ensuring equitable access to renewable energy benefits, regardless of socioeconomic status, is crucial for social justice.

Balancing Environmental and Social Considerations:

1. **Integrated Planning**: Comprehensive planning that

considers environmental and social factors is essential to minimize negative impacts and maximize benefits.

2. **Site Selection**: Choosing suitable locations for renewable energy projects is important to minimize environmental and social conflicts.

3. **Regulation and Standards**: Robust regulations, environmental impact assessments, and social safeguards are essential to guide responsible renewable energy development.

4. **Mitigation Measures**: Implementing mitigation measures, such as habitat restoration, noise reduction, and community compensation, can help address negative impacts.

5. **Transparency and Accountability**: Transparent communication, accountability mechanisms, and monitoring of impacts are vital for maintaining trust with local communities.

Striking a balance between achieving renewable energy goals and addressing environmental and social concerns requires collaboration among stakeholders, adherence to best practices, and a commitment to sustainability. Renewable energy projects that take into account both environmental and social considerations can contribute to a more resilient and equitable energy future.

Tidal, Wave, and Ocean Thermal Energy Conversion (OTEC)

Tidal energy, wave energy, and ocean thermal energy conversion (OTEC) are three distinct types of marine renewable energy technologies that harness the power of the ocean to generate electricity. Each technology utilizes different principles and mechanisms to convert oceanic energy into usable energy. Here's an overview of each:

Tidal Energy:

Tidal energy exploits the gravitational forces between the Earth, the Moon, and the Sun to generate electricity from the rise and fall of tides. Tidal energy can be harnessed using two main methods:

1. **Tidal Range Energy**: This involves constructing tidal barrages or tidal fences across estuaries or coastal areas. As tides rise and fall, water flows through turbines in the barriers, generating electricity. Tidal range energy is reliable and predictable but requires suitable tidal variations.

2. **Tidal Current Energy**: Tidal currents are the horizontal movement of water caused by the tides. Underwater turbines or underwater kite-like structures can be placed in areas with strong tidal currents to capture kinetic energy and generate electricity.

Wave Energy:

Wave energy is generated by the motion of ocean waves. Devices placed on the ocean surface or submerged in the water capture the energy of the moving waves and convert it into electricity. Wave energy technologies include:

1. **Point Absorbers**: These devices float on the surface and use hydraulic systems to convert the vertical motion of waves into pressurized fluid, which then drives a turbine to generate electricity.
2. **Oscillating Water Columns**: These are semi-submerged structures with an air chamber. As waves enter the chamber, the trapped air is compressed, and the resulting air flow drives a turbine.
3. **Attenuators and Overtopping Devices**: These devices are placed perpendicular to the direction of wave movement and capture the energy of wave crests and troughs. They use the differential height of water to generate power.

Ocean Thermal Energy Conversion (OTEC):

OTEC utilizes the temperature difference between warm surface waters and cold deep waters in the ocean to generate electricity. The process involves the following steps:

1. **Warm Surface Water**: Warm surface water is used to vaporize a working fluid (usually ammonia) with a low boiling point.
2. **Vaporization and Turbine**: The vaporized fluid drives a turbine, generating electricity.
3. **Cold Deep Water**: Cold deep water is used to condense the vaporized fluid back into a liquid state, completing the cycle.

OTEC requires a significant temperature difference between surface and deep waters, typically found in tropical regions.

Advantages and Challenges:

Advantages:

- Tidal energy offers predictable and reliable power generation due to the consistent and cyclic nature of tides.
- Wave energy can provide constant energy production, as waves are prevalent in various oceanic regions.
- OTEC can provide continuous and baseload power, especially in tropical areas with the required temperature difference.

Challenges:

- High initial costs and technical challenges in constructing and maintaining ocean-based energy installations.
- Environmental impacts, including effects on marine ecosystems, navigation, and coastal dynamics.
- Variability in resource availability and energy output due to weather conditions and geographical constraints.

Tidal, wave, and OTEC technologies have the potential to contribute to global renewable energy goals, but their commercial viability and widespread adoption depend on advancements in technology, cost reductions, and effective environmental management.

Challenges and Potential of Ocean Energy

Ocean energy, encompassing tidal, wave, and ocean thermal energy conversion (OTEC), holds significant promise as a renewable energy source. However, it also faces several challenges that need to be addressed for its full potential to be realized. Here are some challenges and the potential of ocean energy:

Challenges:

1. **Technical Challenges**: Developing efficient and durable technologies for harnessing ocean energy, such as robust materials that can withstand harsh marine conditions, is a significant challenge.
2. **High Capital Costs**: The upfront costs of building and installing ocean energy infrastructure, including tidal barrages, wave energy converters, and OTEC plants, are often high compared to other renewable energy sources.
3. **Resource Variability**: The availability of ocean energy resources can be inconsistent due to factors like changing tidal patterns, variable wave heights, and regional temperature differences required for OTEC.
4. **Environmental Impact**: Installing ocean energy devices and infrastructure can potentially impact marine ecosystems, habitats, and navigation. Mitigating these impacts while pursuing sustainable energy generation is essential.
5. **Energy Transport**: Transmitting electricity generated from offshore installations to onshore grids can pose challenges due to long distances and underwater

cabling requirements.

6. **Deployment and Maintenance**: Maintaining and repairing ocean energy installations in remote and harsh marine environments can be logistically complex and expensive.

7. **Permitting and Regulatory Hurdles**: Obtaining permits and approvals for ocean energy projects can be challenging due to environmental, navigational, and regulatory considerations.

8. **Competing Land and Ocean Uses**: Ocean areas suitable for energy generation may also be used for fishing, shipping, recreation, and conservation, leading to conflicts over resource allocation.

Potential:

1. **Abundant and Predictable Resource**: Ocean energy sources, such as tides and waves, are driven by natural phenomena like gravitational forces and wind patterns, making them relatively predictable and abundant.

2. **High Energy Density**: Ocean energy has a high energy density compared to some other renewable sources, meaning a smaller installation can produce a significant amount of energy.

3. **Reduced Emissions**: Ocean energy can contribute to reducing greenhouse gas emissions and dependence on fossil fuels, thus helping mitigate climate change.

4. **Local Energy Generation**: Offshore installations can provide localized energy generation, reducing the need for long-distance electricity transmission and improving energy security.

5. **Job Creation**: The development, installation, and maintenance of ocean energy projects can create jobs in coastal communities, contributing to local economies.

6. **Complementary to Other Renewables**: Ocean energy can complement other renewable sources like wind and solar, providing a more balanced and reliable energy mix.

7. **Long-Term Potential**: With technological advancements and experience gained from pilot projects, the cost-effectiveness and efficiency of ocean energy technologies could improve over time.

To unlock the full potential of ocean energy, collaborative efforts are required among governments, industries, research institutions, and communities. Investments in research and development, advancements in technology, policy support, and environmental considerations will be vital in addressing the challenges and harnessing the benefits of this promising renewable energy source.

Harnessing the Power of the Seas

"Harnessing the power of the seas" refers to the utilization of the ocean's energy resources, including tidal, wave, and ocean thermal energy, to generate electricity and contribute to the global transition to renewable energy sources. This concept involves capturing the immense energy present in oceanic processes and converting it into usable power. Here's a summary of how different ocean energy sources can be harnessed:

1. **Tidal Energy**: Tidal energy is generated by the gravitational forces between the Earth, Moon, and Sun, causing the rise and fall of tides. Tidal energy can be harnessed using tidal range energy (using barrages or fences) or tidal current energy (using underwater turbines or other devices) to generate electricity.

2. **Wave Energy**: Wave energy is produced by the movement of waves on the ocean's surface. Devices like point absorbers, oscillating water columns, and attenuators capture this motion and convert it into electricity through various mechanisms.

3. **Ocean Thermal Energy Conversion (OTEC)**: OTEC utilizes the temperature difference between warm surface waters and cold deep waters to generate electricity. This process involves using a working fluid to drive a turbine by vaporizing and condensing the fluid.

By tapping into these ocean energy sources, we can benefit from:

- **Predictable Power Generation**: Ocean energy sources, especially tidal energy, offer predictable patterns that make energy generation more reliable than some other renewable sources like wind and solar.

- **High Energy Density**: Ocean energy has a high energy density, meaning a relatively small installation can produce a substantial amount of power.
- **Renewable Energy Generation**: Ocean energy is renewable and sustainable, helping reduce greenhouse gas emissions and dependence on fossil fuels.
- **Local Energy Production**: Offshore installations can provide localized energy generation, improving energy security and reducing the need for long-distance electricity transmission.
- **Job Creation and Economic Development**: The development, construction, and maintenance of ocean energy projects can create jobs and stimulate local economies, especially in coastal communities.

However, the concept of harnessing the power of the seas also involves addressing challenges like high upfront costs, technical complexities, environmental impacts, and regulatory hurdles. Collaborative efforts among governments, industries, research institutions, and communities are essential to navigate these challenges and unlock the full potential of ocean energy as a sustainable and impactful source of clean electricity.

Battery Technologies and Grid-Scale Storage

Battery technologies and grid-scale storage play a crucial role in the integration of renewable energy sources into the electricity grid. These technologies help address the intermittency of renewable energy generation and ensure a stable and reliable power supply. Here's an overview of battery technologies and grid-scale storage:

Battery Technologies:

Battery technologies store electrical energy in chemical form and convert it back into electricity when needed. They are used for a wide range of applications, from small-scale residential systems to large grid-scale installations. Some common battery technologies include:

1. **Lithium-Ion Batteries**: Widely used for portable electronics and electric vehicles, lithium-ion batteries have high energy density and are suitable for various applications, including grid-scale storage.
2. **Flow Batteries**: These batteries use two electrolyte solutions separated by a membrane. They can be scaled up for grid-scale applications and offer advantages in terms of capacity and longevity.
3. **Sodium-Sulfur Batteries**: These high-temperature batteries are used for grid-scale applications due to their high energy density and efficiency.
4. **Lead-Acid Batteries**: Traditional lead-acid batteries are relatively low-cost options used in off-grid and

backup power systems.

5. **Solid-State Batteries**: Under development, solid-state batteries promise higher energy density, improved safety, and longer lifespans compared to current technologies.

6. **Redox Flow Batteries**: Similar to flow batteries, redox flow batteries use vanadium-based electrolytes and can be used for grid-scale energy storage.

Grid-Scale Storage:

Grid-scale storage involves large-scale energy storage systems that are integrated into the electricity grid to provide stability, manage peak demand, and balance supply and demand fluctuations. Grid-scale storage technologies include:

1. **Battery Energy Storage Systems (BESS)**: Large banks of batteries can store excess electricity during times of low demand and release it during peak demand or when renewable sources are not generating.

2. **Pumped Hydro Storage**: This involves using surplus electricity to pump water from a lower reservoir to an upper reservoir. During peak demand, the water is released through turbines to generate electricity.

3. **Compressed Air Energy Storage (CAES)**: Excess electricity is used to compress air and store it in underground caverns. When needed, the compressed air is released to drive turbines and generate electricity.

4. **Flywheel Energy Storage**: High-speed rotating flywheels store kinetic energy and release it as electricity when needed.

5. **Hybrid Systems**: Combining different storage technologies can optimize energy storage capacity, efficiency, and response times.

Advantages:

- **Balancing Supply and Demand**: Grid-scale storage helps match electricity supply with demand, reducing the need for fossil fuel-based peaker plants during high-demand periods.
- **Integration of Renewables**: Storage systems smooth out the intermittent nature of renewable energy sources, ensuring a reliable and stable power supply.
- **Enhanced Grid Stability**: Storage can provide grid stability by responding quickly to fluctuations in supply and demand, helping to prevent blackouts and voltage instability.
- **Energy Market Support**: Grid-scale storage can participate in energy markets by buying electricity during low-demand periods and selling it during high-demand periods, benefiting both consumers and providers.

Challenges:

- **Cost**: The upfront costs of grid-scale storage systems can be significant, and the levelized cost of energy storage is a crucial consideration.
- **Efficiency**: Energy losses during charging and discharging reduce the overall efficiency of storage systems.
- **Environmental Impact**: Some battery technologies involve the use of rare and potentially environmentally harmful materials.
- **Regulatory and Market Barriers**: Regulatory frameworks and market structures need to evolve to accommodate the integration of storage into the grid.

Grid-scale storage and advanced battery technologies are essential components of a resilient, flexible, and sustainable energy system. As technology advances and costs decrease, these solutions are expected to play an increasingly vital role in supporting the transition to a clean energy future.

Pumped Hydro Storage and Compressed Air Energy Storage

Pumped Hydro Storage and Compressed Air Energy Storage (CAES) are two prominent grid-scale energy storage technologies that play a critical role in stabilizing electricity grids, managing peak demand, and integrating renewable energy sources. Let's delve into each technology:

Pumped Hydro Storage:

Pumped Hydro Storage is one of the most mature and widely used forms of grid-scale energy storage. It involves using the potential energy of water by moving it between two reservoirs located at different elevations. Here's how it works:

1. **Charging (Off-Peak Periods):**
   - During periods of low electricity demand or when renewable sources generate excess power, the surplus electricity is used to pump water from a lower reservoir to an upper reservoir.
   - This process converts electrical energy into gravitational potential energy, storing it as the water is held at a higher elevation.
2. **Discharging (Peak Demand or Supply Shortages):**
   - When electricity demand is high or renewable energy sources are generating less power, the stored water from the upper reservoir is released.
   - As the water flows downhill, it passes

through turbines that generate electricity through hydroelectric generators.

Advantages:

- High efficiency: Pumped hydro storage systems can achieve high round-trip efficiency, typically around 70-85%.
- Large storage capacity: These systems can store a substantial amount of energy, making them suitable for grid balancing over extended periods.
- Long lifespan: Pumped hydro facilities have a long operational lifespan, often exceeding 50 years.
- Environmental benefits: They produce no direct emissions and have a minimal impact on the environment if sited properly.
- Quick response: Pumped hydro storage can respond rapidly to changes in demand, providing grid stability.

Compressed Air Energy Storage (CAES):

CAES is another form of grid-scale energy storage that utilizes compressed air to store and release energy. There are two main types of CAES: adiabatic and diabatic.

1. **Diabatic CAES**:
   - During off-peak periods, excess electricity is used to compress air and store it in an underground cavern or storage vessel.
   - When electricity demand rises, the compressed air is released and mixed with natural gas in a combustion chamber.
   - The combustion produces high-temperature gases that expand and drive a turbine to generate electricity.
2. **Adiabatic CAES**:
   - Similar to diabatic CAES, excess electricity is used to compress air and store it in

> underground caverns.
> - However, instead of combustion, the released air is heated using an external heat source before passing through an expansion turbine.

Advantages:

- Scalability: CAES systems can be scaled to provide large storage capacities, making them suitable for grid stability.
- Grid integration: CAES can help balance electricity supply and demand, especially during peak periods.
- Energy-efficient: When utilizing waste heat or renewable heat sources, adiabatic CAES can achieve high round-trip efficiency.
- Location flexibility: CAES facilities can be located near existing gas infrastructure and geological formations suitable for air storage.

Challenges and Considerations for Both Technologies:

- Site availability: Both pumped hydro and CAES require specific geological and topographical conditions, which might limit their deployment in some regions.
- Environmental impact: Site selection is crucial to minimize environmental disruption and land use conflicts.
- Infrastructure costs: Building the necessary infrastructure, such as reservoirs, turbines, and caverns, can be capital-intensive.
- Efficiency losses: Energy losses during storage and conversion processes can impact the overall efficiency of these systems.
- Regulatory considerations: Permitting and regulatory processes can be complex and time-consuming for both technologies.

Both pumped hydro storage and CAES are essential components

of the energy storage mix, contributing to grid stability and supporting the transition to a more sustainable and renewable energy future.

Balancing Supply and Demand with Energy Storage

Balancing supply and demand is a critical aspect of maintaining a stable and reliable electricity grid. Energy storage technologies, such as batteries, pumped hydro storage, compressed air energy storage (CAES), and more, play a crucial role in achieving this balance. Here's how energy storage helps balance supply and demand:

1. Storing Excess Energy:

During periods of low electricity demand or when renewable energy sources (such as solar and wind) are generating more power than needed, energy storage systems can store the excess energy for later use. This excess energy is captured and stored, preventing it from going to waste.

2. Shifting Energy:

Energy storage allows for the shifting of energy from times of high supply to times of high demand. For instance, energy generated during the day when solar panels are producing at their peak can be stored and discharged during the evening when demand is higher.

3. Peak Demand Management:

Energy storage systems can be deployed to manage peak demand periods. When electricity demand surges during specific times of the day, such as early evening, energy storage can release stored power to supplement the grid and prevent strain on power generation resources.

4. Grid Stability:

Renewable energy sources, like solar and wind, are variable in nature, leading to fluctuations in electricity generation. Energy storage systems can provide rapid response capabilities, injecting power into the grid during sudden drops in renewable generation to maintain grid stability.

5. Frequency Regulation:

Energy storage can help regulate the frequency of the grid. Any mismatch between electricity supply and demand affects the grid's frequency, which needs to be maintained within a specific range. Energy storage systems can provide or absorb power to balance the frequency.

6. Ancillary Services:

Energy storage can provide ancillary services to the grid, such as voltage regulation, reactive power support, and black start capabilities. These services enhance the stability and reliability of the grid.

7. Integration of Renewable Energy:

Energy storage enables a smoother integration of variable renewable energy sources into the grid. It mitigates the challenges associated with the intermittent nature of wind and solar power.

8. Avoiding Grid Congestion:

In areas where transmission lines are congested, energy storage can help alleviate congestion by storing excess energy and releasing it when the grid can handle the load.

By effectively balancing supply and demand, energy storage technologies contribute to a more resilient, flexible, and efficient electricity grid. They enable the grid to accommodate a higher share of renewable energy, reduce the reliance on fossil fuels, and enhance overall energy system reliability.

Smart Grids and Demand Response

Smart grids and demand response are innovative approaches that enhance the efficiency, reliability, and sustainability of the electricity grid by incorporating advanced technologies, real-time communication, and consumer engagement. These concepts play a crucial role in optimizing energy use, managing peak demand, and integrating renewable energy sources. Here's an overview of smart grids and demand response:

Smart Grids:

A smart grid is an intelligent electricity network that uses advanced technologies, sensors, communication systems, and data analytics to improve the management of electricity generation, distribution, and consumption. Key features of smart grids include:

1. **Advanced Metering Infrastructure (AMI)**: Smart meters provide real-time data on energy consumption, allowing consumers and utilities to monitor and manage energy use more effectively.
2. **Two-Way Communication**: Smart grids enable bidirectional communication between utilities and consumers. This enables real-time information exchange, allowing both sides to make informed decisions.
3. **Distributed Energy Resources (DERs)**: Smart grids integrate and manage a variety of DERs, including solar panels, wind turbines, energy storage systems, and electric vehicles.
4. **Grid Automation and Control**: Automation

technologies enhance grid reliability by quickly identifying and addressing issues, such as power outages and equipment failures.

5. **Predictive Analytics**: Data analytics and artificial intelligence are used to forecast energy demand, optimize grid operations, and enhance energy efficiency.

6. **Demand Side Management**: Smart grids enable demand response programs that encourage consumers to adjust their energy use based on real-time pricing or grid conditions.

Demand Response:

Demand response is a strategy that involves adjusting electricity consumption in response to changes in energy prices or grid conditions. It encourages consumers to reduce or shift their energy use during peak demand periods to relieve stress on the grid and avoid the need to use expensive and less environmentally friendly peaker plants. Demand response mechanisms include:

1. **Time-of-Use (TOU) Pricing**: Consumers are charged different rates for electricity based on the time of day. Higher rates during peak periods incentivize energy use during off-peak hours.

2. **Critical Peak Pricing (CPP)**: In this approach, electricity rates are increased significantly during critical peak demand periods to encourage consumers to reduce consumption.

3. **Load Curtailment Programs**: Consumers voluntarily reduce energy use during peak demand events in exchange for incentives or rebates.

4. **Automated Demand Response**: Smart technologies and devices are used to automate energy reduction strategies, such as adjusting thermostat settings or turning off non-essential equipment.

Benefits:

- **Grid Stability**: Smart grids and demand response enhance grid stability by reducing the risk of overloads and blackouts during peak demand periods.
- **Energy Efficiency**: Consumers can optimize their energy use, leading to reduced energy waste and lower utility bills.
- **Integration of Renewables**: Demand response can help manage the variability of renewable energy sources, such as solar and wind, by shifting consumption to align with their availability.
- **Cost Savings**: By reducing peak demand and the need for expensive peaker plants, utilities can save on operational costs, which can ultimately benefit consumers.
- **Environmental Impact**: Smart grids and demand response contribute to reducing greenhouse gas emissions by reducing the need for fossil fuel-based power generation.

Challenges and Considerations:

- **Technological Integration**: Implementing smart grids requires significant investments in technology, infrastructure, and communication systems.
- **Consumer Engagement**: Demand response relies on consumer participation and willingness to modify energy consumption habits.
- **Privacy and Data Security**: Smart meters and data analytics raise concerns about consumer privacy and the secure handling of energy consumption data.
- **Regulatory Framework**: Regulatory support is necessary to incentivize utilities and consumers to participate in demand response programs.
- **Education and Awareness**: Consumer education is

vital to ensure that people understand the benefits of demand response and how to participate effectively.

Smart grids and demand response are essential tools for optimizing energy systems, accommodating renewable energy growth, and enhancing grid reliability. They empower consumers to play an active role in the energy transition and promote a more sustainable and resilient energy future.

Microgrids and Localized Energy Systems

Microgrids and localized energy systems are innovative approaches to energy distribution that focus on creating more resilient, efficient, and sustainable energy networks on a smaller scale. These systems provide a way to generate, store, and distribute energy within specific areas, communities, or facilities, often incorporating renewable energy sources and advanced technologies. Here's an overview of microgrids and localized energy systems:

Microgrids:

A microgrid is a localized energy system that can operate independently or in conjunction with the main power grid. It includes distributed energy resources (DERs) such as solar panels, wind turbines, energy storage systems, and backup generators. Microgrids are managed through advanced control systems and can function autonomously to provide electricity to specific areas or facilities. Key characteristics of microgrids include:

1. **Energy Independence**: Microgrids can operate autonomously during grid outages by relying on their own energy generation and storage capabilities.
2. **Integration of Renewables**: Microgrids often incorporate renewable energy sources, reducing reliance on fossil fuels and contributing to sustainability goals.
3. **Reliability and Resilience**: Microgrids enhance energy

reliability and resilience by providing backup power during grid failures or disruptions.

4. **Island Mode Operation**: Microgrids can isolate themselves from the main grid and operate in island mode, ensuring continuous power supply to critical facilities.

5. **Demand Management**: Advanced control systems optimize energy use within the microgrid, managing demand and supply to reduce costs and improve efficiency.

6. **Grid Support**: Microgrids can support the main grid by providing excess energy or helping manage peak demand.

Localized Energy Systems:

Localized energy systems involve generating and utilizing energy within specific communities, neighborhoods, or even individual buildings. These systems prioritize local energy generation, storage, and consumption to reduce dependence on centralized power sources. Key aspects of localized energy systems include:

1. **Decentralization**: Localized energy systems distribute energy generation across multiple points, reducing the need for long-distance transmission.

2. **Community Engagement**: Communities play an active role in managing and benefiting from localized energy systems, fostering a sense of ownership and sustainability.

3. **Renewable Integration**: These systems often prioritize renewable energy sources like solar and wind to minimize environmental impact.

4. **Energy Efficiency**: Localized systems can implement energy-efficient technologies and practices tailored to local needs.

5. **Energy Security**: By diversifying energy sources and

reducing reliance on external grids, localized systems enhance energy security.

6. **Economic Benefits**: Communities can potentially save on energy costs by generating and consuming energy locally.

Benefits and Considerations:

Benefits:

- **Resilience**: Microgrids enhance energy resilience, reducing the impact of grid failures and outages.
- **Renewable Integration**: Both microgrids and localized systems facilitate the integration of renewable energy sources, contributing to sustainability goals.
- **Efficiency**: Localized systems promote efficient energy use and distribution, reducing transmission losses.
- **Community Empowerment**: These systems empower communities to take control of their energy supply and consumption.
- **Economic Savings**: By generating and using energy locally, communities can potentially save on energy costs over the long term.

Considerations:

- **Technical Complexity**: Designing, implementing, and managing microgrids and localized systems require technical expertise and investment.
- **Regulatory Challenges**: Regulatory frameworks and policies may need to adapt to accommodate the integration of localized energy systems.
- **Interconnection**: Ensuring seamless interaction between microgrids, localized systems, and the main grid can be complex.
- **Investment**: Initial capital investment might be higher, but long-term savings and benefits can outweigh the costs.

Microgrids and localized energy systems offer the potential to transform how communities generate, consume, and manage energy. They contribute to energy independence, sustainability, and resilience while empowering communities to actively participate in the transition to a cleaner and more efficient energy future.

Overcoming Intermittency and Ensuring Reliability

Overcoming intermittency and ensuring reliability in energy systems, particularly those reliant on renewable sources like solar and wind, is a fundamental challenge in the transition to a cleaner and more sustainable energy future. Intermittency refers to the unpredictable and varying nature of renewable energy generation due to factors like weather conditions. Here are some strategies to address intermittency and ensure reliability in energy systems:

1. Energy Storage:

Energy storage technologies, such as batteries, pumped hydro storage, and thermal storage, store excess energy when renewable sources are producing more than needed and release it when demand is higher. These systems can bridge the gap between variable renewable generation and demand, providing consistent and reliable power supply.

2. Hybrid Systems:

Combining different renewable energy sources can mitigate the impact of intermittency. For example, a hybrid system could integrate solar and wind power, as they often have complementary generation patterns – sunny days might have less wind, while windy days might be cloudy.

3. Demand Response:

Demand response programs incentivize consumers to adjust

their energy consumption based on grid conditions and pricing signals. By shifting energy-intensive tasks to times when renewable generation is high, demand response can help match supply with demand.

4. Flexible Generation:

Having a mix of generation sources, including dispatchable sources like natural gas plants, can provide flexible capacity that can be adjusted to meet demand during periods of low renewable generation.

5. Forecasting and Predictive Analytics:

Advanced weather forecasting and predictive analytics can provide more accurate predictions of renewable energy generation. This enables grid operators to better anticipate changes and adjust power generation and consumption accordingly.

6. Grid Integration and Balancing Tools:

Sophisticated grid management tools can balance supply and demand across regions. High-voltage transmission lines and grid interconnections can also help distribute energy from areas with excess generation to those with higher demand.

7. Energy Market Designs:

Designing energy markets that value flexibility, storage, and demand response can incentivize investments in technologies that help manage intermittency and ensure grid reliability.

8. Ancillary Services:

Ancillary services like frequency regulation, voltage control, and reserves play a crucial role in grid stability. Energy storage and flexible resources can provide these services to maintain grid reliability.

9. Technological Innovation:

Continued advancements in renewable energy technologies, energy storage, grid management systems, and smart grid technologies will improve the ability to manage intermittency and enhance reliability.

10. Grid Planning and Infrastructure Investment:

Planning for renewable energy integration and investing in necessary grid upgrades and infrastructure can ensure that the grid can accommodate higher levels of intermittent renewable generation.

Addressing intermittency and ensuring energy system reliability is a multi-faceted effort that requires a combination of technological, regulatory, and market-driven solutions. As renewable energy technologies and grid management strategies continue to evolve, the energy industry can increasingly mitigate the challenges associated with intermittency and pave the way for a more sustainable and reliable energy future.

Renewable Energy Economics and Cost Competitiveness

Renewable energy economics and cost competitiveness have undergone significant transformations over the years, making renewable sources increasingly competitive with traditional fossil fuels. Several factors have contributed to this shift, including technological advancements, economies of scale, policy support, and growing environmental concerns. Here's an overview of renewable energy economics and cost competitiveness:

1. Cost Reductions:

Technological Advancements: Advances in renewable energy technologies, such as solar panels and wind turbines, have led to increased efficiency and lower manufacturing costs.

Scale Effects: As renewable energy deployment has grown, economies of scale have kicked in, reducing the cost per unit of energy generated.

Learning Curve: With increased production and experience, the industry has benefited from the learning curve effect, leading to better processes and cost reductions.

2. Policy Support:

Subsidies and Incentives: Governments worldwide have provided financial incentives, tax credits, grants, and feed-in tariffs to encourage the adoption of renewable energy technologies.

Renewable Portfolio Standards (RPS): Mandates that require a certain percentage of energy to come from renewable sources have driven demand and deployment.

3. **Market Forces**:

Competitive Bidding: Auctions and competitive bidding processes have encouraged developers to offer renewable energy projects at increasingly lower prices.

Investor Interest: The growing interest from investors and financiers in sustainable investments has led to lower financing costs for renewable projects.

4. **Grid Parity and Levelized Cost of Energy (LCOE)**:

As renewable energy costs have decreased, they have reached grid parity – the point at which the cost of renewable energy is equal to or lower than that of conventional energy sources.

LCOE is a metric used to compare the costs of different energy sources over their lifetime. It considers factors like capital costs, operating costs, and expected energy output. The decreasing LCOE of renewables has made them more cost-competitive.

5. **Externalities and Environmental Concerns**:

Renewable energy sources produce little to no greenhouse gas emissions and have minimal air and water pollution compared to fossil fuels. As concerns about climate change and environmental impacts have grown, the value of these external benefits has increased, making renewables more attractive.

6. **Dynamic Markets**:

Energy markets are dynamic, and prices of fossil fuels can fluctuate due to geopolitical factors, supply disruptions, and changing demand. Renewable energy, being domestically produced, can provide energy security and price stability.

7. Innovation and Research:

Continued research and innovation in renewable energy technologies have the potential to drive further cost reductions, making renewables even more economically competitive.

Challenges and Considerations:

- **Intermittency and Storage**: While costs of renewables have dropped, addressing the intermittency of sources like solar and wind through energy storage still presents challenges.
- **Initial Investment**: While renewable energy costs have decreased significantly, there are still upfront capital costs associated with building new infrastructure.
- **Policy Uncertainty**: Changes in government policies and incentives can affect the cost competitiveness of renewable projects.
- **Infrastructure and Grid Integration**: Upgrading grid infrastructure to accommodate higher levels of renewables can pose challenges.

Overall, the economics of renewable energy have evolved favorably, enabling renewables to compete with conventional energy sources. Continued advancements, innovation, supportive policies, and a global shift towards sustainability are expected to further enhance the cost competitiveness of renewable energy and accelerate the transition to a clean energy future.

Government Incentives and Subsidies

Government incentives and subsidies play a crucial role in promoting the adoption and growth of renewable energy technologies. These policy measures aim to reduce the upfront costs of renewable energy projects, create market demand, stimulate investment, and accelerate the transition to a cleaner and more sustainable energy system. Here are some common types of government incentives and subsidies for renewable energy:

1. Feed-in Tariffs (FiTs):

Feed-in tariffs guarantee a fixed payment rate for renewable energy producers, usually above the market price, over a specific contract period. This incentivizes the development of renewable energy projects by ensuring a predictable and attractive return on investment.

2. Production Tax Credits (PTCs) and Investment Tax Credits (ITCs):

These are tax incentives provided to renewable energy project developers and investors. PTCs are based on the amount of renewable energy generated, while ITCs offer a percentage reduction in project costs. These incentives effectively lower the cost of renewable energy projects and encourage private investment.

3. Renewable Portfolio Standards (RPS) and Renewable Energy Targets:

RPS require utilities to obtain a specific percentage of their energy from renewable sources. Governments set these targets to drive demand for renewable energy and create a market for renewable energy technologies.

4. Grants and Subsidies:

Direct grants and subsidies provide financial support to renewable energy projects, helping to offset initial capital costs and improve project economics.

5. Tax Incentives and Exemptions:

Governments often offer tax breaks, exemptions, or deductions to renewable energy projects, reducing their tax burden and improving financial viability.

6. Net Metering and Feed-in Policies:

Net metering allows energy consumers with their own renewable energy systems to sell excess electricity back to the grid. Feed-in policies enable energy producers to sell their generated energy to the grid at favorable rates.

7. Green Certificates and Tradable Renewable Energy Credits (RECs):

Green certificates and RECs represent the environmental attributes of renewable energy generation. Governments may require utilities to purchase a certain number of these certificates, creating additional revenue streams for renewable energy projects.

8. Low-Interest Loans and Financing Programs:

Governments may offer low-interest loans or favorable financing terms to renewable energy projects to reduce the financial burden on developers and investors.

9. Research and Development (R&D) Funding:

Funding for R&D in renewable energy technologies helps drive innovation, improve efficiency, and lower costs.

10. Import Duty Exemptions:

Exemptions on import duties for renewable energy equipment can significantly reduce the upfront costs of installing renewable energy systems.

Benefits:

- **Market Growth**: Incentives create demand for renewable technologies, leading to increased market growth and job creation.
- **Economic Stimulus**: Investments in renewable energy projects stimulate local economies through job creation, technology deployment, and infrastructure development.
- **Reduced Emissions**: Incentives help reduce greenhouse gas emissions by promoting the use of cleaner energy sources.
- **Technological Innovation**: Support for R&D encourages innovation, leading to advancements in renewable energy technologies.
- **Energy Security**: Diversifying the energy mix with renewables enhances energy security by reducing dependence on fossil fuels.

Challenges and Considerations:

- **Budget Constraints**: Governments must balance the cost of incentives with other spending priorities.
- **Policy Stability**: Frequent changes to incentive programs can create uncertainty and impact investor confidence.
- **Market Distortions**: Poorly designed incentives can distort energy markets or lead to inefficiencies.
- **Equity Concerns**: Incentives should be structured to ensure equitable distribution of benefits and avoid disproportionately benefiting certain groups.

Well-designed and targeted government incentives and subsidies can drive the adoption of renewable energy technologies, accelerate the energy transition, and contribute to a more sustainable energy future.

International Agreements and Climate Goals

International agreements and climate goals are vital tools in addressing global environmental challenges, particularly the urgent need to reduce greenhouse gas emissions and combat climate change. These agreements provide a framework for global cooperation and coordinated action to mitigate the impacts of climate change. Here are some key international agreements and climate goals:

1. United Nations Framework Convention on Climate Change (UNFCCC):

The UNFCCC, established in 1992, is the foundational international treaty aimed at addressing climate change. Its objective is to prevent "dangerous anthropogenic interference with the climate system." The UNFCCC sets the stage for negotiations on more specific agreements.

2. Kyoto Protocol:

Adopted in 1997 and in force from 2005 to 2020, the Kyoto Protocol legally bound developed countries to specific emissions reduction targets. It introduced the concept of emissions trading and Clean Development Mechanism (CDM) projects to promote emission reductions in developing countries.

3. Paris Agreement:

The Paris Agreement, adopted in 2015 and ratified by nearly all countries, is a landmark global accord within the UNFCCC framework. Its main goal is to limit global warming to well below 2 degrees Celsius above pre-industrial levels, aiming for 1.5 degrees Celsius. It sets a framework for countries to submit nationally determined contributions (NDCs) outlining their emissions reduction targets and strategies.

4. Nationally Determined Contributions (NDCs):

Under the Paris Agreement, countries submit NDCs detailing their intended contributions to emissions reduction. These contributions vary based on each country's national circumstances, capabilities, and ambitions.

5. Sustainable Development Goals (SDGs):

The SDGs, established by the United Nations in 2015, include Goal 13: "Take urgent action to combat climate change and its impacts." This goal emphasizes the integration of climate action into broader sustainable development efforts.

6. Net-Zero Emissions Goals:

Many countries and entities have set net-zero emissions goals, committing to balance the amount of greenhouse gases emitted with the amount removed from the atmosphere. Achieving net-zero emissions is a critical step in limiting global temperature rise.

7. International Energy Agency (IEA) Goals:

The IEA has called for a "Net Zero by 2050" pathway to achieve global net-zero emissions by 2050 and limit warming to 1.5 degrees Celsius.

8. G20 Climate Commitments:

The G20, a group of major economies, has made commitments to address climate change through various means, including clean energy deployment, finance, and technology transfer.

Benefits and Challenges:

Benefits:

- **Global Cooperation**: International agreements facilitate cooperation and shared responsibilities among nations.
- **Emission Reductions**: Climate goals drive emissions reductions, contributing to mitigating the impacts of climate change.
- **Technology Transfer**: Agreements encourage the transfer of clean technologies to developing countries.
- **Enhanced Resilience**: Climate goals promote measures to enhance climate resilience and adaptation.

Challenges:

- **Differing Priorities**: Countries have varying economic, social, and developmental priorities, which can complicate negotiations.
- **Implementation Challenges**: Ensuring countries meet their commitments and targets requires effective monitoring and enforcement mechanisms.
- **Equity and Fairness**: Disparities between developed and developing countries raise concerns about equity in burden sharing.
- **Political Will**: International agreements depend on political will and cooperation, which can be influenced by changing leadership and national interests.

International agreements and climate goals provide a framework for collective action to address the urgent challenges of climate change. They encourage countries to work together, set ambitious targets, and implement policies that will help transition to a sustainable and low-carbon future.

Life Cycle Analysis of Renewable Energy Sources

Life Cycle Analysis (LCA) is a comprehensive approach used to assess the environmental impacts of a product, process, or technology throughout its entire life cycle, from raw material extraction to disposal. LCA is a valuable tool for evaluating the environmental sustainability of renewable energy sources and comparing them to conventional energy sources. Here's how LCA is applied to renewable energy sources:

1. Raw Material Extraction and Manufacturing:

The life cycle of a renewable energy source begins with the extraction of raw materials for manufacturing. LCA assesses the environmental impacts associated with mining, refining, and processing these materials. For example, in solar panels, the production of photovoltaic cells requires materials like silicon, metals, and chemicals.

2. Manufacturing and Assembly:

LCA evaluates the energy consumption, emissions, and resource use during the manufacturing and assembly of renewable energy technologies. This includes processes such as fabricating solar panels, manufacturing wind turbine components, or producing biofuel feedstock.

3. Installation and Construction:

The construction and installation phase involves the transportation, assembly, and setup of renewable energy systems. LCA considers the environmental impacts of transporting and installing wind turbines, solar arrays, and other infrastructure.

4. Operation and Maintenance:

LCA assesses the ongoing energy production and operational phase of renewable energy systems. Factors like energy output, maintenance requirements, and system efficiency are considered.

5. End-of-Life:

The final stage of the life cycle involves decommissioning and disposal. LCA examines the environmental impacts of dismantling, recycling, and managing waste from renewable energy systems.

Key Environmental Impact Categories in LCA:

LCA analyzes various environmental impact categories, which may include:

- **Greenhouse Gas Emissions**: Assessing the emissions of carbon dioxide (CO_2), methane (CH_4), and other greenhouse gases over the life cycle.
- **Energy Use**: Measuring the energy consumption throughout the life cycle, including direct and indirect energy inputs.
- **Resource Depletion**: Evaluating the consumption of non-renewable resources, such as minerals and fossil fuels, and its impact on resource availability.
- **Air and Water Pollution**: Analyzing emissions of pollutants like sulfur dioxide (SO_2), nitrogen oxides (NOx), and particulate matter that can affect air and water quality.
- **Land Use**: Examining the land area required for renewable energy systems and potential impacts on ecosystems and biodiversity.

Comparing Renewable Energy Sources:

LCA allows for comparisons between different renewable energy sources and conventional energy sources. It considers the entire life cycle, which can lead to more accurate assessments of the environmental benefits and drawbacks of each option.

Challenges and Considerations:

- **Data Availability**: Gathering accurate data for all life cycle stages can be challenging and impact the reliability of LCA results.
- **Boundary Definition**: Defining the boundaries of the life cycle (e.g., from cradle to grave or cradle to gate) can affect the results.
- **Geographical Variability**: Environmental impacts can vary based on location, resource availability, and energy mix.
- **Assumptions**: LCA requires assumptions for factors like energy use, recycling rates, and technology advancements, which can impact results.

Life Cycle Analysis is a valuable tool for assessing the environmental performance of renewable energy sources, guiding policy decisions, and driving improvements in technology and processes to minimize environmental impacts.

Mitigating Environmental Footprint

Mitigating the environmental footprint is essential for promoting sustainable development, reducing ecological impact, and combating climate change. It involves minimizing the negative environmental consequences of human activities across various sectors. Here are strategies to mitigate the environmental footprint:

1. Transition to Renewable Energy:

Shifting from fossil fuels to renewable energy sources such as solar, wind, hydro, and geothermal can significantly reduce greenhouse gas emissions and air pollution associated with energy production.

2. Energy Efficiency:

Implementing energy-efficient technologies and practices in industries, buildings, transportation, and appliances reduces energy consumption and lowers emissions.

3. Sustainable Transportation:

Promoting public transportation, cycling, walking, and electric vehicles can reduce emissions from the transportation sector.

4. Circular Economy:

Adopting a circular economy approach focuses on reducing waste, reusing materials, and recycling to minimize resource consumption and landfill waste.

5. Green Building Practices:

Constructing energy-efficient, eco-friendly buildings with efficient heating, cooling, lighting, and water systems can lower energy consumption and emissions.

6. Reforestation and Afforestation:

Planting trees and restoring natural habitats helps sequester carbon dioxide, combat deforestation, and support biodiversity.

7. Conservation and Biodiversity Protection:

Preserving natural ecosystems and protecting biodiversity contributes to ecological balance and the health of the planet.

8. Sustainable Agriculture:

Implementing sustainable farming practices such as agroforestry, organic farming, and precision agriculture reduces land degradation, chemical use, and emissions.

9. Water Conservation:

Efficient water management practices, reduction of water waste, and protection of water resources are critical for sustainability.

10. Waste Reduction and Management:

Implementing effective waste management strategies, promoting recycling and composting, and reducing single-use plastics are crucial.

11. Sustainable Consumption and Lifestyle:

Promote conscious consumer behavior, reduce overconsumption, and choose products with lower environmental impacts.

12. Carbon Pricing and Regulations:

Implementing carbon pricing mechanisms and regulations that limit emissions encourage businesses to transition to cleaner practices.

13. Innovation and Technology:

Investing in research and innovation for green technologies, including carbon capture and storage, helps mitigate environmental impacts.

14. Education and Awareness:

Raising public awareness about environmental issues and encouraging sustainable behaviors can drive positive change.

15. International Cooperation:

Collaboration among governments, industries, NGOs, and international organizations is crucial for addressing global environmental challenges.

Benefits and Challenges:

Benefits:

- Improved Environmental Quality: Mitigation efforts lead to cleaner air, water, and ecosystems.
- Economic Opportunities: Green technologies and sustainable practices can create new economic opportunities and jobs.
- Climate Resilience: Mitigation reduces vulnerability to climate change impacts.

Challenges:

- Economic Costs: Implementing sustainable practices may initially require investments and changes in business models.
- Policy Complexity: Effective policies require coordination and collaboration across sectors and stakeholders.
- Behavior Change: Encouraging behavior change and overcoming resistance to new practices can be

challenging.

Mitigating the environmental footprint is an urgent and collective responsibility. By adopting sustainable practices, policies, and technologies, we can safeguard the planet's health and ensure a more sustainable future for current and future generations.

Renewable Energy's Role in Carbon Emission Reduction

Renewable energy plays a pivotal role in reducing carbon emissions and addressing the global challenge of climate change. Here's how renewable energy contributes to carbon emission reduction:

1. Low or Zero Emissions:

Renewable energy sources, such as solar, wind, hydro, and geothermal, produce little to no direct greenhouse gas emissions during their operation. Unlike fossil fuels, which release carbon dioxide (CO_2) and other pollutants when burned, renewables generate electricity without emitting harmful gases.

2. Substitution of Fossil Fuels:

Renewable energy technologies can replace fossil fuels for electricity generation, heating, cooling, and transportation. By substituting coal, oil, and natural gas with cleaner energy sources, renewable energy reduces CO_2 emissions and air pollutants.

3. Energy Transition:

Transitioning from fossil fuel-based energy systems to renewable energy systems helps decarbonize the economy. As renewables become a larger share of the energy mix, carbon emissions from energy production decrease.

4. Electricity Sector Transformation:

The electricity sector is a significant contributor to carbon emissions. By replacing coal-fired power plants with wind and solar farms, emissions are reduced while clean energy capacity is increased.

5. Decentralized Energy Generation:

Renewable energy allows for distributed energy generation, reducing the need for long-distance energy transportation and transmission losses. This enhances energy efficiency and reduces emissions associated with energy transmission.

6. Flexibility and Storage:

Advancements in energy storage technologies enable the integration of renewable energy sources into the grid, even when the sun isn't shining or the wind isn't blowing. Energy storage improves grid stability, minimizes the use of backup fossil fuel generators, and optimizes renewable energy utilization.

7. Electrification of Transport:

Renewable energy can power electric vehicles (EVs), reducing emissions from the transportation sector. This is especially effective when paired with an increase in renewable energy adoption in the electricity sector.

8. Carbon Intensity Reduction:

Renewable energy technologies have lower carbon intensity compared to fossil fuels. The carbon intensity of an energy source refers to the amount of CO_2 emitted per unit of energy generated. This reduction contributes to overall emissions reduction.

9. Climate Goals and Agreements:

Many countries have committed to reducing their carbon emissions in alignment with international agreements like the

Paris Agreement. Renewable energy is a central strategy to achieving these emission reduction targets.

10. Public and Private Initiatives:

Growing public awareness and private sector initiatives have led to increased investments in renewable energy projects and technologies, accelerating the transition away from fossil fuels.

Benefits and Challenges:

Benefits:

- **Climate Mitigation**: Renewable energy significantly reduces carbon emissions, mitigating climate change impacts.
- **Air Quality Improvement**: Transitioning to renewables improves air quality by reducing emissions of harmful pollutants.
- **Energy Security**: Diversifying the energy mix with renewables enhances energy security and reduces dependence on fossil fuel imports.
- **Economic Opportunities**: The renewable energy sector creates jobs, stimulates innovation, and drives economic growth.

Challenges:

- **Intermittency**: Some renewable sources are intermittent, requiring energy storage and grid flexibility to ensure reliable power supply.
- **Infrastructure and Investment**: Transitioning to renewables requires infrastructure investment, policy support, and financing.
- **Market Competition**: Fossil fuel industries may resist the transition due to economic interests.

The expansion of renewable energy is essential for achieving a sustainable and low-carbon future. By reducing carbon emissions, renewables contribute to global efforts to limit global warming and mitigate the impacts of climate change.

Technological Advancements on the Horizon

Technological advancements in various fields are continuously shaping the future and have the potential to revolutionize industries, improve quality of life, and address global challenges. Here are some technological advancements on the horizon that could have a significant impact:

1. **Artificial Intelligence (AI) and Machine Learning**:

AI and machine learning are rapidly advancing, enabling machines to learn from data and perform tasks that traditionally required human intelligence. AI has applications in various sectors, including healthcare diagnostics, autonomous vehicles, personalized marketing, and climate modeling.

2. **Quantum Computing**:

Quantum computers use quantum bits (qubits) to process information in ways that classical computers cannot. Quantum computing could revolutionize fields like cryptography, optimization, and material science, potentially solving complex problems faster and more efficiently.

3. **Renewable Energy Advancements**:

Technological innovations in renewable energy are driving efficiency improvements and cost reductions. This includes advanced solar cell designs, more efficient wind turbine blades, enhanced energy storage solutions, and grid integration technologies.

4. Electric and Autonomous Vehicles:

Electric vehicles (EVs) are becoming more affordable and have longer ranges, contributing to the shift away from internal combustion engines. Autonomous vehicles are also advancing, with potential benefits for safety, efficiency, and transportation accessibility.

5. Biotechnology and Genomics:

Advances in biotechnology, including gene editing techniques like CRISPR-Cas9, have the potential to revolutionize healthcare, agriculture, and environmental conservation by enabling precise genetic modifications.

6. Space Exploration and Colonization:

Technological advancements are enabling ambitious plans for space exploration and even human colonization of other planets. Private companies are developing technologies for reusable rockets, lunar exploration, and Mars missions.

7. Synthetic Biology:

Synthetic biology involves designing and engineering biological systems for useful purposes. It has applications in creating biofuels, bioplastics, new pharmaceuticals, and even custom organisms for various industrial processes.

8. Advanced Materials:

Materials science is advancing with the development of new materials that have unique properties. This includes materials for energy storage, lightweight and strong construction materials, and materials with advanced thermal and electronic properties.

9. Healthcare and Telemedicine:

Advancements in healthcare technologies, including

telemedicine, wearable health devices, and personalized medicine based on genetic information, are transforming healthcare delivery and improving patient outcomes.

10. **Blockchain and Decentralized Technologies**:

Blockchain technology has potential applications beyond cryptocurrencies, such as secure supply chain management, digital identity verification, and transparent voting systems.

11. **Climate and Environmental Technologies**:

Technologies for carbon capture and storage, sustainable agriculture practices, efficient waste management, and renewable energy are crucial for addressing climate change and preserving the environment.

12. **Neuroscience and Brain-Computer Interfaces**:

Advances in neuroscience are leading to breakthroughs in understanding the brain and developing brain-computer interfaces. This has potential applications in treating neurological disorders and enabling direct communication between the brain and computers.

These technological advancements hold the promise to reshape industries, societies, and the way we interact with the world. However, they also come with ethical, regulatory, and social considerations that need to be carefully navigated to ensure that they are used for the greater benefit of humanity.

Overcoming Infrastructure and Scalability Challenges

Overcoming infrastructure and scalability challenges is crucial for the successful deployment of technological advancements, especially in sectors like renewable energy, transportation, and emerging technologies. These challenges can hinder the widespread adoption of new technologies and limit their potential impact. Here are strategies to address infrastructure and scalability challenges:

1. **Investment in Infrastructure**:

Governments, private companies, and international organizations need to invest in building the necessary infrastructure to support new technologies. This includes energy grids for renewable energy, charging networks for electric vehicles, and high-speed internet for digital technologies.

2. **Public-Private Partnerships**:

Collaboration between governments and private sector entities can accelerate infrastructure development. Public-private partnerships can leverage resources, expertise, and funding to address scalability challenges.

3. **Regulatory Frameworks**:

Clear and supportive regulatory frameworks are essential for the adoption of new technologies. Regulations should encourage innovation while ensuring safety, security, and compliance with environmental and social standards.

4. **Standardization**:

Developing and adhering to industry standards can facilitate interoperability and compatibility among different technologies. Standardization reduces complexity and accelerates deployment.

5. **Incentives and Subsidies**:

Government incentives and subsidies can encourage the adoption of new technologies. These can include tax incentives, grants, and subsidies to reduce the initial cost barriers for individuals and businesses.

6. **Demonstration Projects**:

Demonstration projects showcase the feasibility and benefits of new technologies on a smaller scale before full-scale implementation. These projects can attract investment and build public confidence.

7. **Research and Development**:

Investments in research and development drive innovation and address technical challenges that may hinder scalability. This includes improving efficiency, reliability, and performance.

8. **Pilot Programs**:

Pilot programs allow for testing and refining technologies in real-world scenarios. They provide insights into challenges that need to be addressed before broader implementation.

9. **Scalability Planning**:

Early planning for scalability is critical. Technologies should be designed with scalability in mind, considering factors such as manufacturing capacity, supply chains, and distribution networks.

10. Educational and Training Programs:

Providing education and training to professionals and the workforce in new technologies ensures that there are skilled individuals capable of implementing, maintaining, and operating these technologies.

11. Public Awareness and Acceptance:

Educating the public about the benefits and importance of new technologies can build support and create demand, driving the need for infrastructure development.

12. Flexible Funding Models:

Innovative funding models, such as pay-as-you-go or leasing options, can make new technologies more accessible to a broader range of users.

Benefits and Considerations:

Benefits:

- Accelerated Adoption: Addressing infrastructure and scalability challenges leads to quicker adoption of technologies.
- Economic Growth: Investment in infrastructure and technology deployment can stimulate economic growth and job creation.
- Sustainability: Technologies like renewable energy and sustainable transportation contribute to environmental sustainability and reduce carbon emissions.

Considerations:

- Cost: Building infrastructure and scaling up technologies can involve significant upfront costs.
- Interconnected Challenges: Overcoming scalability challenges often requires addressing multiple

interconnected factors.

- Public Perception: Public perception and concerns about the impact of new technologies can influence their adoption.
- Technical Complexity: Some technologies may require specialized skills and expertise for installation, operation, and maintenance.

By implementing a combination of strategies and fostering collaboration among stakeholders, infrastructure and scalability challenges can be effectively addressed, allowing new technologies to reach their full potential and drive positive change.

Global Vision for a Sustainable Energy Landscape

A global vision for a sustainable energy landscape involves a comprehensive and coordinated effort to transition to a cleaner, more efficient, and equitable energy system. Such a vision aligns with international commitments, addresses climate change, fosters economic development, and ensures energy security. Here are key elements of a global vision for a sustainable energy landscape:

1. Decarbonization and Renewable Energy Adoption:

The primary goal is to rapidly decarbonize the energy sector by replacing fossil fuels with renewable energy sources such as solar, wind, hydro, and geothermal. This transition minimizes greenhouse gas emissions and contributes to limiting global temperature rise.

2. Energy Efficiency and Conservation:

Promote energy efficiency across all sectors, from transportation and industry to buildings and agriculture. Emphasize the importance of reducing energy waste through better technologies, practices, and behavioral changes.

3. Electrification of Multiple Sectors:

Expand the use of electricity in sectors traditionally reliant on fossil fuels, such as transportation and heating. Electric vehicles, electrified public transportation, and electric heating systems contribute to emissions reduction.

4. Smart and Resilient Energy Infrastructure:

Invest in modern, decentralized, and resilient energy infrastructure that integrates renewable energy, energy storage, and advanced grid technologies. Smart grids enable efficient distribution and management of electricity.

5. Universal Access to Energy:

Ensure universal access to reliable, affordable, and clean energy. Address energy poverty by providing clean energy solutions to underserved communities, improving livelihoods, and promoting social equity.

6. Carbon Capture and Removal:

Implement technologies for carbon capture, utilization, and removal to offset remaining emissions and achieve net-zero carbon emissions. This includes direct air capture, reforestation, and soil carbon sequestration.

7. Collaboration and International Cooperation:

Promote global cooperation and knowledge sharing to accelerate the adoption of sustainable energy technologies. International agreements and partnerships facilitate the exchange of expertise and resources.

8. Research and Innovation:

Invest in research and innovation to develop new technologies that enhance energy efficiency, storage, and conversion. Encourage interdisciplinary collaboration to address complex energy challenges.

9. Circular Economy Principles:

Promote a circular economy by minimizing waste and maximizing the use of resources throughout the energy value chain. Recycle materials, reduce environmental impacts, and extend the lifecycle of products.

10. **Education and Awareness**:

Raise public awareness about the importance of sustainable energy and its benefits for climate, health, and economic development. Empower individuals and communities to make informed energy choices.

11. **Policy and Regulation**:

Implement supportive policies, regulations, and incentives that promote the adoption of sustainable energy technologies. Price carbon emissions, phase out fossil fuel subsidies, and provide incentives for renewable energy deployment.

12. **Resilience and Adaptation**:

Consider climate resilience in energy planning and infrastructure development. Design energy systems that can withstand and recover from extreme weather events and other climate-related challenges.

Benefits and Challenges:

Benefits:

- Mitigating Climate Change: A sustainable energy landscape is key to achieving global climate goals and limiting global temperature rise.
- Improved Air and Water Quality: Transitioning to clean energy sources reduces air pollution and protects water resources.
- Energy Security: Diversification of energy sources and decentralized systems enhance energy security and resilience.
- Economic Growth: The transition to sustainable energy creates jobs, stimulates innovation, and fosters economic growth.

Challenges:

- Infrastructure Investment: Building the necessary infrastructure for sustainable energy requires significant investment and planning.
- Technological and Regulatory Barriers: Some technologies are still in the developmental stages and face regulatory challenges.
- Transition Costs: The upfront costs of transitioning from fossil fuels to renewable energy can be a barrier for some regions.
- Social Equity: Ensuring that the benefits of a sustainable energy landscape are accessible to all requires addressing social and economic disparities.

A global vision for a sustainable energy landscape requires international collaboration, strong leadership, and a commitment to making bold choices for the benefit of current and future generations.

Solar Success Stories: Innovations and Implementation

Solar energy has seen remarkable success stories in both technological innovations and widespread implementation. These success stories showcase the potential and impact of solar energy in various sectors. Here are some examples:

1. Grid-Scale Solar Power Plants:

Solar power plants have been developed on a massive scale, delivering clean energy to communities and industries. The Noor Ouarzazate Solar Complex in Morocco, for instance, is one of the world's largest solar power plants, harnessing solar energy through concentrated solar power (CSP) technology. It contributes to Morocco's energy security and reduces carbon emissions.

2. Residential and Commercial Solar Rooftop Installations:

Innovations in photovoltaic technology and financing models have enabled widespread adoption of solar panels on residential and commercial rooftops. Germany is a leader in this regard, with its Feed-in Tariff policy spurring a surge in solar installations, transforming energy landscapes and empowering individuals to generate their own power.

3. Floating Solar Farms:

Floating solar farms have gained attention as an innovative way to utilize water bodies for renewable energy production. For example, China's "Three Gorges Dam" has a large floating solar installation, benefiting from the water's cooling effect on solar panels and reducing water evaporation.

4. Solar-Powered Desalination:

Solar energy is being used to power desalination plants, addressing both water scarcity and energy challenges. In countries like Saudi Arabia and the United Arab Emirates, solar-powered desalination facilities provide clean drinking water by using solar energy to drive the desalination process.

5. Solar-Powered Electric Vehicles (EVs):

Solar panels integrated into EVs have the potential to extend their range and reduce reliance on grid charging. Companies like Lightyear and Sono Motors are developing solar-powered electric vehicles that can charge from sunlight, promoting sustainable transportation.

6. Solar-Powered Microgrids:

Solar microgrids are providing reliable and clean energy to remote and off-grid communities. In India, for instance, the state of Rajasthan has implemented solar-powered microgrids to bring electricity to villages that were previously without access to power.

7. Solar-Powered Airports and Infrastructure:

Solar installations at airports and other large infrastructure projects demonstrate the feasibility of integrating renewable energy into urban environments. The Cochin International Airport in India is fully powered by a solar installation, reducing its carbon footprint.

8. Solar-Powered Healthcare:

Solar energy is improving healthcare access in underserved areas. For example, We Care Solar's "Solar Suitcase" provides solar power to medical clinics in remote regions, ensuring reliable electricity for medical equipment and lighting.

9. **Agricultural Solar Solutions**:

Solar-powered irrigation systems are enhancing agricultural productivity while conserving water. These systems are particularly beneficial in regions with abundant sunlight and limited access to conventional energy sources.

10. **Solar Innovations in Space**:

Innovations in solar energy are extending beyond Earth. Solar panels power satellites, space stations, and exploration missions, enabling scientific research and technological advancements in space.

These solar success stories highlight the diverse applications and benefits of solar energy, from large-scale power generation to improving quality of life in remote areas. Technological innovations and collaborative efforts continue to drive the growth of solar energy as a key player in the global transition to a sustainable energy future.

Wind Energy Triumphs: Projects and Lessons Learned

Wind energy has achieved significant triumphs with the development of large-scale projects and the accumulation of valuable lessons that guide the expansion of this renewable energy source. Here are some examples of wind energy projects and the lessons learned from them:

1. Offshore Wind Farms:

Offshore wind farms have emerged as a major success in wind energy. Projects like the London Array in the UK and the Hornsea Wind Farm also in the UK showcase the potential of harnessing strong and consistent offshore winds. These projects have helped diversify the energy mix, reduce carbon emissions, and create jobs.

Lessons Learned:

- **Technology Advances**: Continuous innovation in offshore wind turbine design and foundation structures has improved efficiency and lowered costs.
- **Grid Integration**: Successful integration of offshore wind farms into the energy grid requires collaboration between energy providers, regulators, and developers.
- **Environmental Impact Assessment**: Comprehensive environmental impact assessments are crucial to mitigate potential effects on marine ecosystems and wildlife.

2. Onshore Wind Farms in Emerging Markets:

Wind energy has made significant strides in emerging markets such as India, Brazil, and China. The Jaisalmer Wind Park in India, for instance, is one of the largest wind energy installations in Asia, contributing to India's renewable energy targets.

Lessons Learned:

- **Local Engagement**: Involving local communities and stakeholders early in the project planning process helps build support and address concerns.
- **Policy Framework**: Clear and stable policy frameworks, including favorable incentives and permitting processes, are essential for attracting investment.
- **Infrastructure Development**: Adequate grid infrastructure and transmission capacity are critical to ensure the successful integration of wind energy into the energy system.

3. Community Wind Projects:

Community wind projects empower local communities to participate in and benefit from renewable energy generation. These projects prioritize community engagement, ownership, and economic development.

Lessons Learned:

- **Local Buy-In**: Engaging the community and involving them in decision-making can lead to greater project acceptance and success.
- **Economic Benefits**: Community wind projects can stimulate local economies by creating jobs, supporting local businesses, and generating revenue.
- **Partnerships**: Collaboration with local governments, organizations, and investors can enhance the viability and sustainability of community wind projects.

4. Hybrid Energy Systems:

Combining wind energy with other renewable sources, such as solar or energy storage, creates hybrid energy systems that offer greater reliability and flexibility.

Lessons Learned:

- **Complementary Resources**: Pairing wind energy with solar or energy storage can provide a more consistent and balanced energy output.
- **Optimized Siting**: Careful site selection is crucial to ensure the synergies between different renewable energy technologies.
- **Grid Management**: Hybrid systems require advanced grid management strategies to optimize energy generation and distribution.

5. Energy Transition Insights:

Wind energy projects have provided valuable insights into the feasibility and challenges of transitioning to renewable energy sources.

Lessons Learned:

- **Technology Costs**: Continuous advancements have significantly lowered the cost of wind energy, making it competitive with conventional sources.
- **Policy Stability**: Clear and consistent energy policies are necessary to attract investment and create a favorable market environment.
- **Integration Challenges**: As the share of wind energy in the grid grows, addressing grid stability, storage, and demand response becomes critical.

These wind energy triumphs and lessons learned demonstrate the viability of wind power as a crucial element of the global shift to sustainable energy systems. Continued innovation, collaborative efforts, and supportive policies will further accelerate the growth and impact of wind energy on a global scale.

Geothermal, Biomass, and Ocean Energy Achievements

Geothermal, biomass, and ocean energy sources have each achieved significant milestones and achievements in their journey towards contributing to the global renewable energy landscape. Here are some examples of accomplishments in each of these areas:

Geothermal Energy Achievements:

1. **Geothermal Power Plants**:
   - The Hellisheiði Power Station in Iceland is one of the largest geothermal power plants in the world. It harnesses geothermal energy for electricity generation and provides both electricity and hot water to the capital city of Reykjavik.

2. **Direct Use Applications**:
   - Geothermal energy is widely used for direct heating and cooling applications. The city of Boise, Idaho, for example, uses geothermal energy to heat buildings, sidewalks, and even a local fish hatchery.

3. **Enhanced Geothermal Systems (EGS)**:
   - EGS projects, which involve creating engineered reservoirs to extract geothermal heat from areas with low natural permeability, are showing promise. The Soultz-sous-Forêts EGS project in France is

an example of efforts to expand geothermal resource availability.

Biomass Energy Achievements:

1. Biofuel Production:
- Advanced biofuels, such as cellulosic ethanol and biodiesel, are being developed from non-food biomass sources like agricultural residues and algae. These fuels have the potential to significantly reduce greenhouse gas emissions compared to conventional fossil fuels.

2. Waste-to-Energy Plants:
- Waste-to-energy plants are converting organic waste into biogas, heat, and electricity. The Reppie Waste-to-Energy Plant in Ethiopia, for instance, is the first waste-to-energy facility in the country and addresses both waste management and energy needs.

3. Cofiring in Power Plants:
- Biomass is being co-fired with coal in existing power plants to reduce emissions and transition to more sustainable energy sources. This approach has been adopted in various countries, including the United States and European nations.

Ocean Energy Achievements:

1. Tidal Energy Installations:
- The MeyGen tidal energy project in Scotland is one of the largest tidal energy arrays in the world. It uses underwater turbines to harness the energy from tidal currents, providing clean and predictable power.

2. Wave Energy Prototypes:

- Wave energy prototypes and test sites have been established to explore the potential of converting wave motion into electricity. The Wave Hub test site in the UK and the Azura wave energy device in the United States are examples of these efforts.

3. **Ocean Thermal Energy Conversion (OTEC):**
   - The OTEC project in Hawaii is exploring the use of temperature differences between warm surface water and cold deep water to generate electricity. OTEC has the potential to provide continuous, baseload power in tropical regions.

These achievements in geothermal, biomass, and ocean energy demonstrate the progress being made toward diversifying the renewable energy mix and unlocking the potential of these resources. As technology continues to advance and these industries grow, these sources have the potential to play a significant role in a sustainable energy future.

Energy Efficiency and Conservation

Energy efficiency and conservation are crucial strategies for reducing energy consumption, mitigating climate change, and ensuring sustainable resource use. They involve optimizing energy use, minimizing waste, and adopting practices and technologies that lead to lower energy consumption. Here's a closer look at energy efficiency and conservation:

Energy Efficiency:

Energy efficiency focuses on achieving the same level of output or service using less energy. It involves improving the efficiency of processes, technologies, and systems to reduce energy waste. Energy-efficient practices have several benefits:

1. **Reduced Energy Consumption**: Using less energy to achieve the same results lowers overall energy consumption and associated costs.

2. **Lower Greenhouse Gas Emissions**: Decreased energy consumption leads to reduced emissions of greenhouse gases like carbon dioxide (CO_2), helping combat climate change.

3. **Cost Savings**: Energy-efficient technologies and practices often result in cost savings for individuals, businesses, and industries through reduced energy bills.

4. **Improved Energy Security**: Reducing energy consumption enhances energy security by reducing dependence on imported fossil fuels.

5. **Enhanced Competitiveness**: Energy-efficient businesses and industries can gain a competitive edge by reducing operational

costs.

Energy Conservation:

Energy conservation involves minimizing energy consumption through behavioral changes and adopting habits that reduce unnecessary energy use. It focuses on eliminating energy waste and making conscious choices to save energy:

1. **Behavioral Changes**: Simple actions like turning off lights when not in use, unplugging electronics, and using natural lighting can lead to significant energy savings.

2. **Efficient Appliances**: Choosing energy-efficient appliances and using them wisely helps reduce energy consumption.

3. **Building Design**: Designing buildings with efficient insulation, windows, and heating/cooling systems reduces energy needs.

4. **Transportation Choices**: Opting for public transportation, carpooling, biking, and driving fuel-efficient vehicles lowers energy consumption in transportation.

5. **Industrial Practices**: Industries can adopt measures such as process optimization, waste heat recovery, and energy audits to conserve energy.

Key Strategies for Energy Efficiency and Conservation:

1. **Technology Adoption**: Invest in energy-efficient technologies such as LED lighting, smart thermostats, and energy-efficient appliances.

2. **Energy Audits**: Conduct energy audits to identify inefficiencies and opportunities for improvement in residential, commercial, and industrial settings.

3. **Policy and Regulations**: Governments can implement energy efficiency standards, labeling programs, and incentives to promote energy-efficient practices.

4. **Education and Awareness**: Educate individuals, communities, and businesses about the importance of energy efficiency and conservation.

5. **Incentives**: Provide financial incentives, tax breaks, and rebates for energy-efficient upgrades and practices.

6. **Building Codes**: Implement strict building codes that require energy-efficient designs and materials.

7. **Research and Innovation**: Continue research and innovation to develop new energy-efficient technologies and practices.

Energy efficiency and conservation are critical components of a sustainable energy future. By adopting these practices on individual, organizational, and societal levels, we can significantly reduce energy waste, lower greenhouse gas emissions, and ensure a more sustainable and resilient energy system.

Community Initiatives and Local Renewable Projects

Community initiatives and local renewable projects play a crucial role in accelerating the adoption of renewable energy, fostering community engagement, and driving sustainable development at the grassroots level. These initiatives empower communities to take ownership of their energy sources and contribute to the transition to clean energy. Here's how community initiatives and local renewable projects make a positive impact:

1. **Ownership and Empowerment**:

Community-driven projects allow local residents to become active participants in the energy transition. When communities have a say in the development and operation of renewable energy projects, they feel a sense of ownership and empowerment.

2. **Economic Benefits**:

Local renewable projects can generate economic benefits for communities. They create jobs, stimulate local economies, and provide a source of revenue for local governments through taxes or revenue-sharing agreements.

3. **Energy Independence**:

Community initiatives contribute to energy independence by reducing reliance on external energy sources, especially fossil fuels. Generating energy locally from renewable sources enhances energy security and resilience.

4. Social Cohesion:

Engaging in renewable energy projects can foster social cohesion and strengthen community bonds. Collaborative decision-making and shared benefits create a sense of unity among community members.

5. Environmental Stewardship:

Local renewable projects promote environmental stewardship by reducing carbon emissions and minimizing the environmental impact of energy production.

6. Education and Awareness:

Community initiatives provide educational opportunities for residents to learn about renewable energy technologies, energy conservation, and environmental sustainability.

7. Model for Change:

Successful local projects can serve as models for neighboring communities, inspiring them to initiate their own renewable energy projects.

Examples of Community Initiatives and Local Renewable Projects:

1. Community Solar Gardens:

These are shared solar installations that allow residents who can't install solar panels on their own property to benefit from solar energy generation. Participants receive credits on their energy bills based on their share of the solar output.

2. Wind Cooperatives:

Community wind cooperatives involve residents collectively owning and operating wind turbines. This model ensures that local communities benefit financially from wind energy production.

3. **Energy Efficiency Programs**:

Community-based energy efficiency initiatives focus on educating residents about energy conservation, offering energy audits, and implementing efficiency measures in homes and businesses.

4. **Bioenergy Projects**:

Communities can develop bioenergy projects that use locally available biomass, such as agricultural waste or wood, to produce heat, electricity, or biogas.

5. **Microgrids**:

Localized microgrids integrate renewable energy sources and energy storage to provide reliable power to a community, even during grid outages.

6. **Renewable Energy Crowdfunding**:

Online platforms allow individuals to contribute funds to support local renewable projects, making it easier for communities to access capital.

Benefits and Challenges:

Benefits:

- Empowerment and Engagement: Communities take an active role in shaping their energy future.
- Economic Development: Local projects create jobs and economic growth.
- Environmental Impact: Reduced greenhouse gas emissions and lower environmental footprint.
- Energy Security: Decreased reliance on external energy sources enhances resilience.
- Education and Awareness: Community initiatives raise awareness about renewable energy and

sustainability.

Challenges:

- Financing: Securing funding can be a challenge for community projects.
- Regulatory Hurdles: Navigating regulations and permits can be complex.
- Technical Expertise: Implementing and maintaining renewable projects may require technical expertise.
- Community Consensus: Building consensus and addressing concerns among community members is important.

Community initiatives and local renewable projects showcase the power of collective action and demonstrate that a sustainable energy future starts at the local level. Through collaboration, innovation, and determination, communities can drive positive change and contribute to a cleaner, more sustainable energy landscape.

The Role of Education in Promoting Sustainable Practices

Education plays a crucial role in promoting sustainable practices and shaping a more environmentally conscious and responsible society. By providing individuals with knowledge, awareness, and the skills needed to make informed choices, education empowers people to adopt behaviors that contribute to a more sustainable future. Here's how education contributes to promoting sustainable practices:

1. **Raising Awareness**:

Education increases awareness about environmental challenges such as climate change, pollution, habitat loss, and resource depletion. It helps individuals understand the consequences of unsustainable practices and the urgency of taking action.

2. **Building Knowledge**:

Education provides a deep understanding of the interconnectedness of ecosystems, the impact of human activities on the environment, and the principles of sustainable resource management.

3. **Fostering Critical Thinking**:

Education encourages critical thinking and analysis. People who are educated about sustainability issues are more likely to question traditional practices and explore alternative solutions.

4. **Encouraging Responsible Consumption**:

Sustainability education highlights the importance of responsible consumption and encourages individuals to make mindful choices that minimize waste, reduce energy use, and support eco-friendly products.

5. **Promoting Renewable Energy Adoption**:

Education about renewable energy sources like solar, wind, and geothermal power helps individuals understand the benefits of clean energy and the role it plays in mitigating climate change.

6. **Empowering Advocacy**:

Educated individuals are more likely to advocate for policy changes, conservation efforts, and sustainable development initiatives within their communities and beyond.

7. **Incorporating Sustainability in Professions**:

Sustainability education equips professionals in various fields, from architecture to business management, with the tools to incorporate sustainable practices into their work.

8. **Encouraging Civic Engagement**:

Education empowers individuals to participate in environmental and social movements, engage in community projects, and support policies that promote sustainability.

9. **Creating Future Leaders**:

Sustainability education nurtures a generation of leaders who are equipped to address complex global challenges, advocate for change, and drive innovation.

10. **Promoting Ethical Values**:

Sustainability education instills ethical values such as respect for the environment, social equity, and intergenerational responsibility.

11. Adapting to Changing Environments:

Education about climate change and adaptation strategies helps communities prepare for the impacts of a changing climate and build resilience.

12. Encouraging Lifelong Learning:

Education is an ongoing process. Encouraging lifelong learning about sustainability ensures that individuals stay informed about evolving environmental issues and solutions.

Integrating Sustainability Education:

- **Formal Education**: Incorporate sustainability concepts into school curricula at all levels, from elementary to higher education.
- **Vocational Training**: Offer training programs that teach sustainable practices in various industries and professions.
- **Community Workshops**: Host workshops and events that educate the public about sustainable practices, waste reduction, and energy efficiency.
- **Online Resources**: Provide accessible online resources, courses, and webinars on sustainability topics.
- **Partnerships**: Collaborate with NGOs, government agencies, and businesses to develop educational initiatives that reach diverse audiences.

By integrating sustainability education into various aspects of society, we can inspire individuals to adopt sustainable practices and collectively work towards a more environmentally friendly, equitable, and resilient world.

Reflection on the Journey

The journey through the various aspects of renewable energy, sustainability, and environmental considerations has been both enlightening and inspiring. From exploring the diverse array of renewable energy sources like solar, wind, geothermal, and ocean energy to understanding the principles of energy efficiency, conservation, and storage, the depth and complexity of our energy systems have become more apparent.

One of the most striking takeaways from this journey is the incredible potential that renewable energy sources hold. The technological advancements, innovative projects, and success stories in different regions of the world showcase the tangible progress we are making towards a cleaner and more sustainable energy landscape. From solar farms to wind turbines, from geothermal power plants to community-driven initiatives, it's evident that a transition to renewable energy is not only feasible but also economically viable and environmentally imperative.

However, this journey has also highlighted the challenges that need to be addressed. The intermittency of some renewable sources, the need for energy storage solutions, the intricacies of grid integration, and the importance of public education and policy support are just a few of the complex aspects that demand attention. Moreover, considering the global context, it's clear that achieving a sustainable energy future requires international collaboration, policy harmonization, and innovative financing mechanisms.

Another significant aspect that emerged is the importance of considering the broader environmental and social impacts of renewable energy technologies. While these sources offer immense benefits in reducing carbon emissions and mitigating climate change, they must be implemented with careful consideration for their potential ecological and community effects. Striking a balance between energy needs, environmental protection, and social equity is a multifaceted challenge that requires interdisciplinary solutions.

As we reflect on this journey, it's evident that renewable energy is not just about technologies; it's about a fundamental shift in how we approach energy production and consumption. It's about recognizing that our energy choices today have a profound impact on the well-being of current and future generations. It's about acknowledging that sustainability is a shared responsibility, encompassing individuals, communities, industries, and governments.

Moving forward, the insights gained from this exploration serve as a reminder of the urgency and potential for positive change. From embracing energy efficiency in our daily lives to supporting large-scale renewable projects, from advocating for policy changes to educating others about sustainable practices, each action contributes to a more sustainable and resilient world. The journey through renewable energy and sustainability is ongoing, but with knowledge, collaboration, and determination, we can navigate the path towards a brighter, cleaner, and more equitable energy future.

The Role of Alternative & Renewable Energy in Shaping a Greener Future

Alternative and renewable energy sources play a pivotal role in shaping a greener and more sustainable future for our planet. As the world faces the urgent challenges of climate change, resource depletion, and environmental degradation, transitioning away from fossil fuels and embracing cleaner energy sources is imperative. Here's how alternative and renewable energy are key to forging a greener future:

1. Reducing Carbon Emissions:

Renewable energy sources, such as solar, wind, and hydroelectric power, emit little to no greenhouse gases during operation. By replacing fossil fuels with renewables, we can significantly reduce carbon emissions, the primary driver of global warming and climate change.

2. Mitigating Air Pollution:

Burning fossil fuels releases pollutants that harm air quality and human health. Shifting to cleaner alternatives minimizes air pollution, reducing respiratory illnesses and improving overall public health.

3. Conserving Natural Resources:

Renewable energy relies on sources that are naturally replenished, such as sunlight, wind, and flowing water. Unlike fossil fuels, these resources are not finite, leading to a more sustainable and balanced use of Earth's resources.

4. Enhancing Energy Security:

Diversifying the energy mix with renewables reduces dependence on imported fossil fuels, enhancing energy security and reducing geopolitical tensions associated with resource extraction.

5. Creating Jobs and Economic Growth:

The renewable energy sector has the potential to create a significant number of jobs in manufacturing, installation, maintenance, and research. Local economic growth is stimulated as communities invest in renewable projects.

6. Promoting Technological Innovation:

The pursuit of renewable energy solutions drives innovation in technology, materials, and processes. This innovation ripples into other sectors, spurring advancements in energy storage, grid management, and efficiency.

7. Empowering Communities:

Local and community-driven renewable energy projects give communities the opportunity to take control of their energy sources, generate revenue, and become more resilient in the face of energy disruptions.

8. Addressing Energy Poverty:

Renewable energy technologies can be deployed to provide electricity to remote and underserved regions, improving living conditions and enhancing opportunities for education, health, and economic development.

9. Strengthening Global Partnerships:

The transition to renewable energy fosters international collaboration as countries share knowledge, technology, and best practices, working together to address global environmental challenges.

10. **Preserving Biodiversity**:

Renewable energy sources have a smaller ecological footprint compared to fossil fuel extraction, minimizing habitat disruption and wildlife displacement.

11. **Promoting Social Equity**:

Renewable energy projects can be designed to prioritize inclusivity, ensuring that benefits are accessible to marginalized communities and not concentrated in a few hands.

12. **Inspiring a Mindset Shift**:

By embracing renewable energy, society fosters a mindset shift towards sustainability, responsible consumption, and a greater appreciation for the interconnectedness of nature.

The path to a greener future involves the collective effort of individuals, communities, governments, and industries to transition towards alternative and renewable energy sources. Embracing these sources not only addresses urgent environmental challenges but also cultivates a more equitable, prosperous, and resilient world for current and future generations.

Inspiring Action and a Sustainable Energy Legacy

Inspiring action and leaving a sustainable energy legacy requires a multi-faceted approach that involves individuals, communities, organizations, and policymakers. By taking concrete steps and fostering a collective commitment to sustainable energy, we can contribute to a greener future and create a positive impact that extends beyond our time. Here's how we can inspire action and leave a sustainable energy legacy:

1. **Raise Awareness and Educate**:

   - Spread awareness about the importance of renewable energy, energy efficiency, and sustainability through education, workshops, and community events.
   - Engage with schools, universities, and community centers to integrate sustainability topics into curricula and promote informed decision-making.

2. **Lead by Example**:

   - Adopt energy-efficient practices in your daily life, such as using LED lights, reducing water usage, and minimizing energy waste.
   - Invest in renewable energy technologies for your home, such as solar panels or energy-efficient appliances, and showcase their benefits to others.

3. **Advocate for Policy Changes**:

   - Join or support advocacy groups that push for policies

that promote renewable energy adoption, energy efficiency standards, and sustainable practices.

- Engage with local, regional, and national policymakers to encourage legislation that advances the renewable energy transition.

4. Promote Community Initiatives:

- Participate in or initiate community-driven renewable energy projects, such as community solar installations, wind cooperatives, or energy-efficient building upgrades.
- Collaborate with local businesses, schools, and organizations to create a shared vision for a sustainable future.

5. Support Renewable Energy Investment:

- Invest in renewable energy projects and companies that align with your values, contributing to the growth of the renewable energy sector.
- Encourage ethical and sustainable investment practices among peers and within financial institutions.

6. Champion Innovation and Research:

- Support research institutions and startups working on innovative renewable energy technologies, energy storage solutions, and grid integration strategies.
- Advocate for increased funding for research and development in the renewable energy field.

7. Engage in Public Discourse:

- Use your voice on social media, blogs, and public forums to discuss the importance of renewable energy and share success stories and lessons learned.
- Engage in constructive conversations about

sustainable practices with family, friends, colleagues, and community members.

8. Volunteer for Environmental Causes:

- Participate in clean-up events, tree planting initiatives, and conservation efforts to demonstrate your commitment to environmental stewardship.
- Volunteer with organizations focused on renewable energy advocacy, climate action, and sustainability.

9. Promote International Collaboration:

- Embrace a global perspective by supporting international agreements and initiatives aimed at advancing renewable energy and addressing climate change.
- Engage in cross-border partnerships that share knowledge and resources to accelerate the transition to sustainable energy.

10. Document and Share Progress:

- Document the progress and impact of your actions, initiatives, and projects to inspire others to follow suit.
- Share success stories, challenges, and lessons learned to encourage a continuous exchange of knowledge.

By taking these actions, you can contribute to a sustainable energy legacy that inspires future generations to prioritize renewable energy, embrace sustainable practices, and work collectively towards a greener, more resilient, and equitable world. Your efforts can create a lasting impact that shapes a positive future for the planet and its inhabitants.

Glossary of Key Terms

Here's a glossary of key terms related to renewable energy, sustainability, and environmental concepts:

Alternative Energy: Energy sources that are alternatives to conventional fossil fuels, such as solar, wind, hydro, geothermal, and biomass.

Biodiversity: The variety of life forms and ecosystems on Earth, including the diversity of species, genes, and habitats.

Carbon Footprint: The total amount of greenhouse gases, mainly carbon dioxide, emitted directly or indirectly by human activities.

Climate Change: Long-term changes in Earth's climate patterns, including global warming caused by human activities such as burning fossil fuels.

Energy Efficiency: The ratio of useful energy output to the energy input, focusing on reducing energy waste and maximizing energy output.

Greenhouse Gas: Gases that trap heat in the Earth's atmosphere, contributing to the greenhouse effect and global warming. Examples include carbon dioxide (CO_2) and methane (CH_4).

Hybrid Energy System: A system that combines different renewable energy sources, such as solar, wind, and storage, to provide a more reliable and continuous energy supply.

Microgrid: A localized energy system that can operate independently or in conjunction with the main grid, often integrating renewable energy sources and storage.

Net Zero Energy: Achieving a balance between the energy consumed and the energy produced or offset, resulting in minimal or zero net energy consumption.

Offshore Wind Farm: Wind turbines located in bodies of water, usually oceans, to harness the strong and consistent winds over water surfaces.

Photovoltaic (PV): Technology that converts sunlight directly into electricity using solar cells made of semiconductor materials.

Renewable Energy: Energy derived from naturally replenished sources, including solar, wind, hydro, geothermal, and biomass, which have minimal environmental impact.

Sustainability: Meeting current needs without compromising the ability of future generations to meet their own needs, while considering environmental, social, and economic factors.

Tidal Energy: Energy harnessed from the gravitational pull of the moon and the sun, causing the periodic rise and fall of sea levels.

Wave Energy: Energy derived from the motion of ocean waves, often harnessed using buoys, oscillating water columns, or other devices.

Wind Turbine: A device that converts the kinetic energy of wind into mechanical energy, which can be used for electricity generation.

Zero-Emission: Producing no greenhouse gas emissions or pollutants during operation. Often used in the context of electric vehicles or renewable energy systems.

This glossary provides a starting point for understanding key terms related to renewable energy and sustainability. As you delve deeper into these topics, you'll encounter more specialized terms that contribute to a comprehensive understanding of our energy systems and their impact on the environment.

Resources and Further Reading

Here are some resources and further reading materials that can provide more in-depth information about renewable energy, sustainability, and related topics:

Books:

1. "Drawdown: The Most Comprehensive Plan Ever Proposed to Reverse Global Warming" by Paul Hawken
2. "The Third Industrial Revolution: How Lateral Power Is Transforming Energy, the Economy, and the World" by Jeremy Rifkin
3. "Renewable Energy: Physics, Engineering, Environmental Impacts, Economics & Planning" by Bent Sørensen
4. "The Green New Deal: Why the Fossil Fuel Civilization Will Collapse by 2028, and the Bold Economic Plan to Save Life on Earth" by Jeremy Rifkin

Websites and Organizations:

1. **Renewable Energy World**: A comprehensive website with news, articles, and resources related to renewable energy technologies and trends. (renewableenergyworld.com)
2. **National Renewable Energy Laboratory (NREL)**: A U.S. Department of Energy laboratory focused on advancing renewable energy and energy efficiency technologies. (nrel.gov)
3. **International Renewable Energy Agency (IRENA)**: An intergovernmental organization that promotes the

adoption of renewable energy on a global scale. (irena.org)

4. **Clean Energy Ministerial (CEM)**: A global forum for sharing best practices and promoting policies and programs that advance clean energy technologies. (cleanenergyministerial.org)

5. **Worldwatch Institute**: A research organization focused on sustainability and global environmental issues. Their reports and publications provide valuable insights. (worldwatch.org)

Journals and Research Publications:

1. **Renewable Energy**: A peer-reviewed journal covering a wide range of renewable energy technologies, research, and applications.

2. **Energy Policy**: A journal that publishes research and analysis on energy policy, economics, and sustainability.

3. **Journal of Cleaner Production**: Focuses on the application of cleaner production principles and sustainable development.

4. **Nature Energy**: A multidisciplinary journal covering all aspects of energy, from policy and economics to technology and sustainability.

Online Courses and MOOCs:

1. **Coursera**: Offers various courses related to renewable energy, sustainability, and environmental topics from universities worldwide. (coursera.org)

2. **edX**: Provides online courses from renowned institutions on renewable energy technologies, energy policy, and sustainable development. (edx.org)

These resources cover a broad spectrum of topics within renewable energy and sustainability, catering to individuals with different levels of expertise and interests. Whether you're looking to deepen your understanding of specific technologies or gain insights into the broader implications of a sustainable energy transition, these resources will be valuable companions on your learning journey.